Lovelock's American Standard of Excellence for Purebred Cattle, Sheep and Swine

by Frank A. Lovelock

with an introduction by Jackson Chambers

This work contains material that was originally published in 1898.

This publication is within the Public Domain.

This edition is reprinted for educational purposes
and in accordance with all applicable Federal Laws.

Introduction Copyright 2018 by Jackson Chambers

Self Reliance Books

Get more historic titles on animal and stock breeding, gardening and old fashioned skills by visiting us at:

http://selfreliancebooks.blogspot.com/

Introduction

I am pleased to present another title in the "Cattle" series.

The work is in the Public Domain and is re-printed here in accordance with Federal Laws.

As with all reprinted books of this age that are intended to perfectly reproduce the original edition, considerable pains and effort had to be undertaken to correct fading and sometimes outright damage to existing proofs of this title. At times, this task is quite monumental, requiring an almost total "rebuilding" of some pages from digital proofs of multiple copies. Despite this, imperfections still sometimes exist in the final proof and may detract from the visual appearance of the text.

I hope you enjoy reading this book as much as I enjoyed making it available to readers again.

Jackson Chambers

PREFACE.

It is now probably some ten or twelve years ago since the editor of this book first became acquainted with "The American Standard of Perfection" for poultry, a work for which his respect and admiration have steadily increased as he has witnessed year by year, at the different poultry shows throughout the country, the skillful work of intelligent expert judges, in determining the merits of competing specimens, by comparing them with the points of excellence laid down in that most reliable authority. Nor has it been only in the show room that the true worth of this perfect poultry guide has been tested, for there is not a careful breeder of pure-bred poultry in America to-day, but who turns with the same regularity for counsel and advice to his "Standard," as the Christian does to his daily guide, the holy Bible. All disputes in competition are settled by the "Standard"; prizes are won by selecting those specimens which approach nearest in excellence to the hundred points of perfection laid down in the "Standard"; buyers purchase those birds which score highest according to the "Standard"; breeders mate their birds according to the requirements of the "Standard," and, inasmuch as the "Standard of Perfection" in poultry is the result of the combined experience and thought of the foremost breeders in the country, so the specimens which in excellence approximate nearest its requirements are considered the best specimens of their breed.

Seeing then, how good a thing was this "Standard of Perfection" as applied to poultry, the editor, who was formerly a breeder and exhibitor of pure-bred live-stock, often thought that a standard of excellence for pure-bred cattle, sheep and swine, if approved by the respective breeders' associations, would be equally as helpful to the thoughful breeder or judge of pure-bred livestock, as those recognized as authorities by the poultry breeders of the United States. At that time it did not occur to the editor of this book to correspond with the secretaries of the different breeders' associations, but, later, after he had commenced his career as expert judge at the Fairs, he gradually obtained the standards of excellence from the aforesaid secretaries, and soon realized that they were of inestimable value in making awards. After having studied them all carefully, the editor is of the opinion that whilst improvements could be made in a few of these, yet any one of them is a vast improvement upon the "rule of thumb" system, (?) for breeding or judging so prevalent throughout the Eastern States. Feeling, therefore, that breeders and judges of pure-bred cattle, sheep and swine would welcome a collection of these standards of excellence in one handy volume, as a guide, and for reference, education and comparison, the undersigned has, after considerable correspondence and study, obtained and systematized almost every standard of excellence ever adopted by any pure-bred live-stock association in America. He trusts that it may become to the breeders and judges of pure-bred live-stock as useful as the "Standard of Perfection" is to the countless breeders of poultry throughout the country.

The undersigned, knowing full well that he could hardly emphasize sufficiently the great need for better judging at Fairs, has taken the liberty of re-publishing from that well-known publication, "The Country Gentleman," an essay upon this topic from the facile pen of the late illustrious Col. F. D. Curtis, which appeared in that excellent weekly, and which should be read by all.

FRANK A. LOVELOCK.

SALEM, VIRGINIA.

A Reform in Judging at Fairs.

EDS. COUNTRY GENTLEMAN—A fair, to fill its full mission, should be educational. When shows simply excite wonder, and only fill gaping mouths with a passing interest, there is not much instruction about them, and very little knowledge is carried home to stimulate improvement and provoke emulation. In all needed reforms, I am not particular about the way it may be done, provided it is done. How shall fairs be made more than an attraction for sight-seeing and the pleasure of meeting each other? It may be urged that there are reasons enough for holding fairs. I grant it, for the past, perhaps, but not for the future. Agriculture must stand in the immediate future upon broader and deeper foundations. There is too big a tide against it to enable it to move with so little power. The propelling force must be stronger, to push our business along in competition with others. The fair must be made more of a factor for instruction and improvement. As now conducted, the "picking up" system of getting judges is too common. This should never be done. None but experts should ever pass upon the merits of goods or animals. By experts I do not mean cranks or ax-grinders. Such men are always out of place, where opinions are asked. There is too much warp in their make-up.

There is a class of experienced and honorable men in every trade, farmers and stock-breeders not excepted. These are the men who should be invited to do the judging, and they should be well paid for it. It is unnecessary to have three experts, as one is ample. The old style judges were generally friends of the powers that be, and while this was no disqualification, it was not an equipment for skill and judgment which specially fitted them for this important and delicate work, nor would it carry much weight with exhibitors and lookers-on.

In the West, where fairs are fairs, there is a rapid tendency towards the one-judge plan, and he a man "known and read of all men" in the special line in which he is called to act. Such judging must be instructive, and far in advance of the awards made upon the "picked up" plan. The judging at some of our State fairs, where we expect a higher grade than at a county fair, is often simply a burlesque. At a State fair recently held, one of the judges, who assumed a degree of arrogance and importance equal to several ordinary men, and wisdom superior to several ordinary owls, did not know anything about swine herd-books or standards, or requirements of associations to constitute thoroughbreds, but he "knew a good hog and one which suited him." Under the dictation and awards of such a judge the exhibitors of all breeds stood back in disgust, and let the thing run. How much could any one learn from such decisions in regard to the characteristics or qualities of any breed or the valuable and best features in breeding?

I would make the exhibit of as great practical value as possible, as this would add to the attractions of the fair. When people found out that teaching by object lessons would take place, and by noted and accepted teachers, they would flock around the rings where stock, was being judged, and the pens, to compare the points and evidences of value. To make all the lessons of the exhibition of animals merely a sight, with printed records to follow, is not the full measure of an agricultural fair of any pretensions. We must get more out of it. The wheels of time now grind too close for so small a return. My idea is that each and every exhibition, especially of live stock, should be a school of thorough object teaching, to be added to all the other and stereotyped impressions. Every animal should be judged by the standard established by the association of the breeders of that class of animals. The standard should be publicly announced, and each animal tried by it, and its superiority over another, or where it may be equal, stated orally; so that those looking on, and the breeders, may see the value of good points and learn how to distinguish them. No one should be allowed to talk back or interfere, but let the judge give reasons for his preferences, and point out the blood markings and perfect features. This kind of judging is no mere dream; its practicability has been demonstrated. This little taste of common sense in judging has made a keen appetite for a full meal. The judges must not be breeders of the same kind of animals, but of other breeds bred for the same purpose. It is difficult to get experts, breeding Jerseys, for instance, who are not identified with some family of this breed, and hence, however high-minded, they would be liable to be accused of leaning toward their favorites. The same criticism might follow with the judges of sheep or swine or horses. Let us have a new track, and see if it will not carry us smoother and better than any old rut. Take the judges for any accepted butter breed from the breeders of butter breeds—from Jerseys to judge Guernseys, Ayrshires to judge Holsteins, and beef breeds in the same way—Short-Horns to judge Herefords—following the same rule with the Polled cattle. The same mismatching should be followed with the breeds of swine, giving, for instance, the Berkshire breeder dominion over the Duroc-Jersey, and the Chester White over the Poland-China, or one judge over each class, classed as large and small—this judge being taken from a class he does not breed.

The fine wool sheep-breeders will oppose a one-judge and an outside man, the strongest, for they have a conceit that a coarse or middle-wool breeder does not know anything about Merinos. Why not? Is there any unfathomable mystery about the breeding or appearance of Merino sheep? The truth is, there ought to be a little more of the middle-wool brains in them, and would it not be a wise thing for the sheep, and the breeders also, to pass in review before the breeders of larger sheep and those bred for a double purpose? There has been too much of the one-idea, or hang-on with the breeders of Merino sheep. They want to get out of it. I should consider it a privilege to have my Merinos judged with the Merino standard by a wide-gauged, middle-wool breeder, and why not the sheep of this last class by a close-texture, fine-staple, big-dewlap, compact-body and hardy-constitution, Merino-educated eye and brain? Sheep must now, of all stock, be bred for utility, and they must stand on their merits while in the hands of the farmer. The protection, trusts and inflation will come when out of the farmer's reach. No one need say that any intelligent breeder, of any class of thoroughbred animals, with the standard of characteristics and the scale of points as a chart, cannot judge fairly and profitably animals bred for the same purpose as his. Such an objection would be a reflection on the intelligence and sound judgment of the breeder. If it may in part hold good, it is a possibility for better results, than with "pick-up" judges, or those expert in their kind, with an inevitable round of dissatisfaction on account of possible prejudice or favoritism.

Kirby Homestead, New York.   F. D. CURTIS.

NOMENCLATURE FOR BULL.

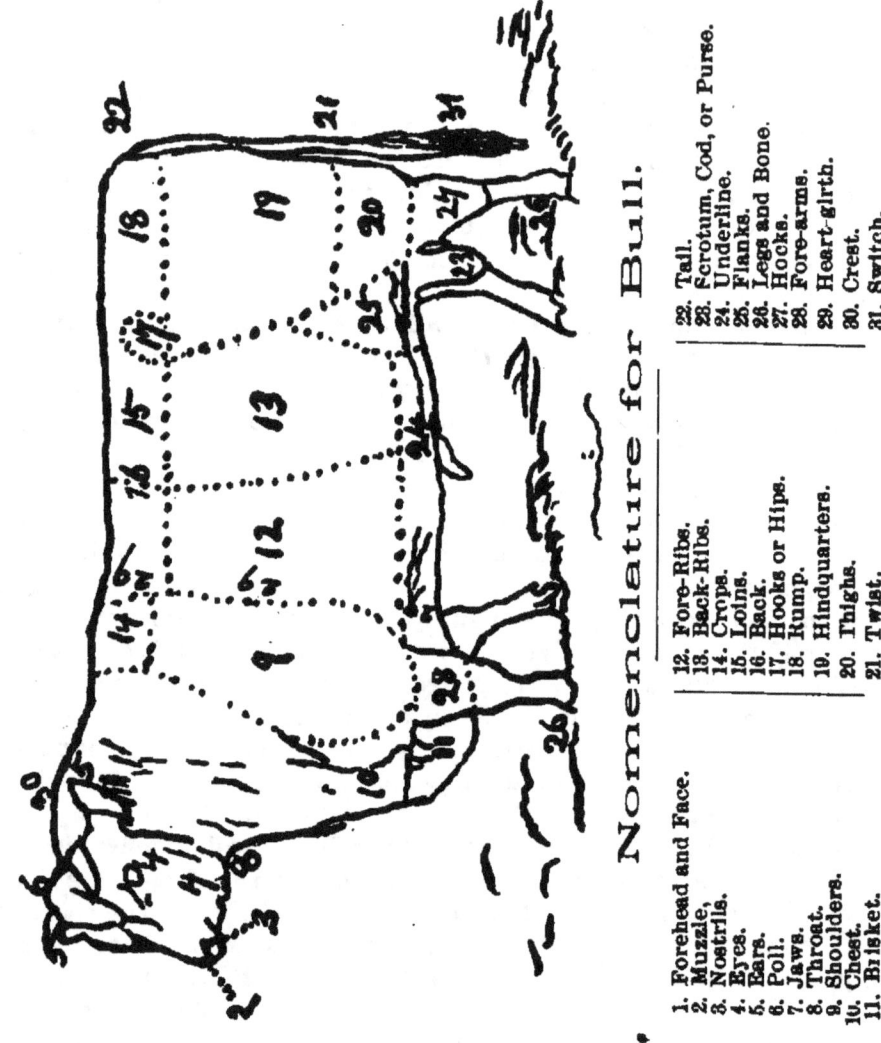

Nomenclature for Bull.

1. Forehead and Face.
2. Muzzle.
3. Nostrils.
4. Eyes.
5. Ears.
6. Poll.
7. Jaws.
8. Throat.
9. Shoulders.
10. Chest.
11. Brisket.
12. Fore-Ribs.
13. Back-Ribs.
14. Crops.
15. Loins.
16. Back.
17. Hooks or Hips.
18. Rump.
19. Hindquarters.
20. Thighs.
21. Twist.
22. Tail.
23. Scrotum, Cod, or Purse.
24. Underline.
25. Flanks.
26. Legs and Bone.
27. Hocks.
28. Fore-arms.
29. Heart-girth.
30. Crest.
31. Switch.

NOMENCLATURE FOR COW.

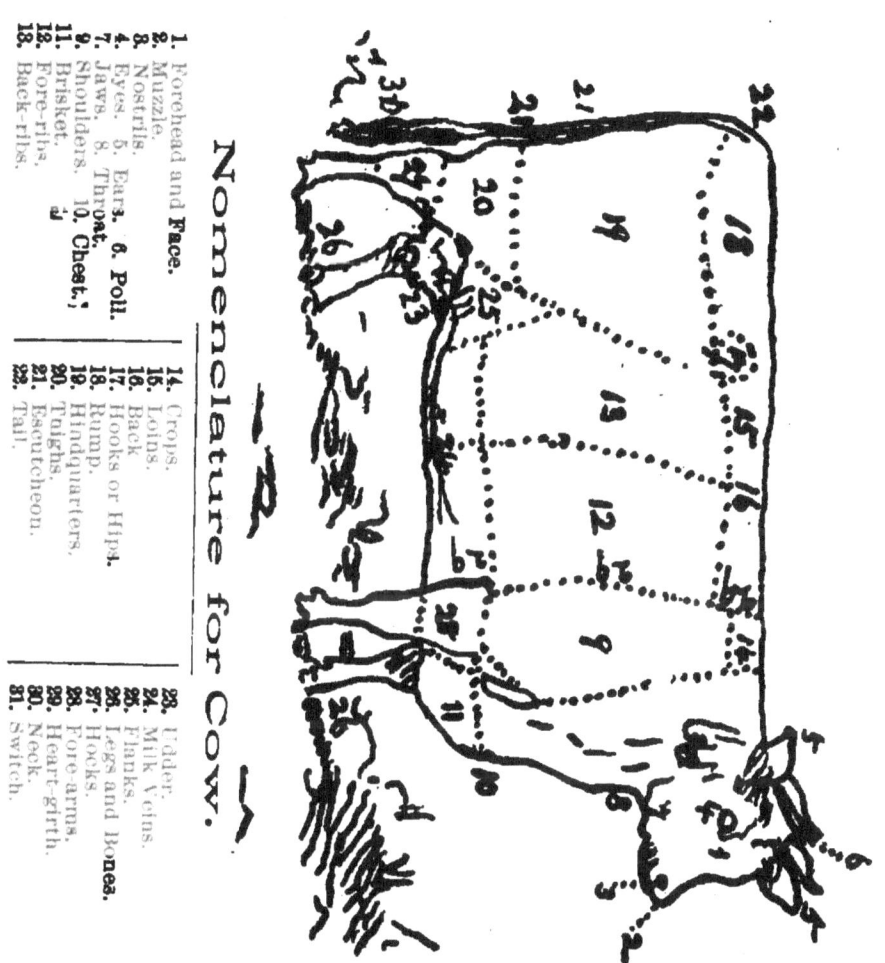

Nomenclature for Cow.

1. Forehead and Face.
2. Muzzle.
3. Nostrils.
4. Eyes. 5. Ears. 6. Poll.
7. Jaws. 8. Throat.
9. Shoulders. 10. Chest.
11. Brisket.
12. Fore-ribs.
13. Back-ribs.
14. Crops.
15. Loins.
16. Back.
17. Hooks or Hips.
18. Rump.
19. Hindquarters.
20. Thighs.
21. Escutcheon.
22. Tail.
23. Udder.
24. Milk Veins.
25. Flanks.
26. Legs and Bones.
27. Hocks.
28. Fore-arms.
29. Heart-girth.
30. Neck.
31. Switch.

ABERDEEN-ANGUS CATTLE.

Standard of Excellence for Aberdeen-Angus Cattle, as adopted by the American Aberdeen-Angus Breeders Association, Thomas McFarlane, Secretary, Harvey, Illinois.

SCALE OF POINTS FOR BULL.

POINTS.	COUNTS.
1. Color,	3
2. Head	10
3. Throat,	3
4. Neck,	3
5. Shoulders,	6
6. Chest,	10
7. Brisket,	4
8. Ribs,	8
9. Back,	10
10. Hindquarters,	8
11. Tail,	3
12. Underline,	4
13. Legs,	4
14. Flesh,	4
15. Skin,	10
16. General Appearance,	10
PERFECTION,	100

When bulls are exhibited with their progeny in a separate class, add 25 counts for progeny.

DETAILED DESCRIPTION.

POINTS.   COUNTS.

1. COLOR.—Black. White is objectionable, except on the underline behind the navel, and there only to a moderate extent; a white cod is most undesirable,   3

2. HEAD.—Forehead broad; face slightly prominent, and tapering toward the nose, muzzle fine; nostrils wide and open; distance from eyes to nostrils of moderate length; eyes mild, full, and expressive, indicative of good disposition; ears of good medium size, well set and well covered with hair; poll well defined, and without any appearance of horns or scurs; jaws clean,   10

3. THROAT.-Clean, without any development of loose flesh underneath, ... 8
4. NECK.—Of medium length, muscular, with moderate crest (which increases with age), spreading out to meet the shoulders, with full neck vein, ... 8
5. SHOULDERS.-Moderately oblique, well covered on the blades and top; with vertebra or backbone slightly above the scapula or shoulder-blades, which should be moderately broad, ... 6
6. CHEST.—Wide and deep; also round and full just back of elbows, ... 10
7. BRISKET.—Deep and moderately projecting from between the legs, and proportionately covered with flesh and fat, 4
8. RIBS.—Well sprung from the backbone, arched and deep, neatly joined to the crops and loins, ... 8
9. BACK.-Broad and straight from crops to hooks; loins strong, hook bones moderate in width, not prominent, and well covered; rumps long, full, level, and rounded neatly into hindquarters, ... 10
10. HINDQUARTERS.—Deep and full, thighs thick and muscular, and in proportion to hindquarters; twist filled out well in its "seam" so as to form an even wide plain between thighs, ... 8
11. TAIL.—Fine, coming neatly out of the body on a line with the back and hanging at right angles to it, . . . 8
12. UNDERLINE.—Straight as nearly as possible, flank deep and full, ... 4
13. LEGS.-Short, straight, and squarely placed, hind legs slightly inclined forward below the hocks; forearm muscular; bones fine and clean, ... 4
14. FLESH.—Even and without patchiness, . . . 4
15. SKIN.-Of moderate thickness and mellow touch, abundantly covered with thick, soft hair. (Much of the thriftiness, feeding properties, and value of the animal depend upon this quality, which is of great weight in the grazier's and butcher's judgment. A good "touch" will compensate for some deficiencies of form. Nothing can compensate for a skin hard and stiff. In raising the skin from the body it should have a substantial, soft, flexible feeling, and when beneath the outspread hand it should move

easily as though resting on a soft cellular substance, which, however, becomes firmer as the animal ripens.

A thin, papery skin is objectionable, especially in a cold climate, ... 10

16. GENERAL APPEARANCE.—Elegant, well bred, and masculine. The walk square, the step quick, and the head up, . 10

PERFECTION, - - - - 100

Purity of blood must be evidenced by registry in the American Aberdeen-Angus Herd Book.

When bulls are exhibited with their progeny in a separate class, add 25 counts for progeny.

POINTS.	SCALE OF POINTS FOR COW.	COUNTS.
1. Color,		2
2. Head,		10
3. Throat,		3
4. Neck,		3
5. Shoulders,		6
6. Chest,		10
7. Brisket,		4
8. Ribs,		8
9. Back,		10
10. Hindquarters,		8
11. Tail,		3
12. Udder,		8
13. Underline,		4
14. Legs,		3
15. Flesh,		8
16. Skin,		10
17. General Appearance,		5
	PERFECTION,	100

In judging heifers omit No. 12. Add 3 counts to No. 15 and 5 counts to No. 17.

DETAILED DESCRIPTION.

POINTS.   COUNTS.

1. COLOR.—Black. White is objectionable, except on the underline, behind the navel, and there only to a moderate extent, ... 2

ABERDEEN-ANGUS CATTLE. 13

2. HEAD.—Forehead moderately broad and slightly indented, tapering toward the nose; muzzle fine; nostrils wide and open; distance from eyes to nostrils of moderate length; eyes full, bright and expressive, indicative of good disposition; ears large, slightly rising upward and well furnished with hair; poll well defined, and without any appearance of horns or scurs; jaws clean, . . . 10

3. THROAT.—Clean, without any development of loose flesh underneath, ... 8

4. NECK.—Of medium length, spreading out to meet the shoulders, with full neck vein, ... 8

5. SHOULDERS.—Moderately oblique, well covered on the blades and top; with vertebra or backbone slightly above the scapula or shoulder-blades, which should be moderately broad, ... 6

6. CHEST.—Wide and deep; round and full just back of elbows. 10

7. BRISKET.—Deep and moderately projecting from between the legs, and proportionately covered with flesh and fat, . 4

8. RIBS.—Well sprung from backbone, arched and deep, neatly joined to the crops and loins, ... 8

9. BACK.—Broad and straight from crops to hooks; loins strong; hook bones moderate in width, not prominent, and well covered; rumps long, full, level, and rounded neatly into hindquarters, ... 10

10. HINDQUARTERS.—Deep and full, thighs thick and muscular, and in proportion with hindquarters; twist filled out well in its "seam," so as to form an even wide plain between thighs, ... 8

11. TAIL.—Fine, coming neatly out of the body on a line with the back and hanging at right angles to it, . . . 8

12. UDDER.—Not fleshy, coming well forward in line with the body, and well up behind; teats squarely placed, well apart and of good size, ... 8

13. UNDERLINE.—Straight as nearly as possible, flank deep and full, ... 4

14. LEGS.—Short, straight, and squarely placed, hind legs slightly inclined forward below the hocks; forearm muscular; bones fine and clean, ... 8

15. FLESH.—Even and without patchiness, . . . 8

16. SKIN.—Of moderate thickness and mellow touch, abund-

antly covered with thick, soft hair. (Much of the thriftiness, feeding properties, and value of the animal depend upon this quality, which is of great weight in the grazier's and butcher's judgment. A good "touch" will compensate for some deficiencies of form. Nothing can compensate for a skin hard and stiff. In raising the skin from the body it should have a substantial, soft, flexible feeling, and when beneath the outspread hand it should move easily as though resting on a soft cellular substance, which, however, becomes firmer as the animal ripens. A thin, papery skin is objectionable, especially in a cold climate), ... 10

17. GENERAL APPEARANCE.—Elegant, well bred, and feminine. The walk square, the step quick, and the head up, . . 5

PERFECTION, - - - - 100

Purity of blood must be evidenced by registry in the American Aberdeen-Angus Herd Book.

In judging heifers omit No. 12. Add 3 counts to No. 15 and 5 counts to No. 17.

AYRSHIRE CATTLE.

Standard of Excellence for Ayrshire Cattle, as adopted by the Ayrshire Breeders Association, C. M. Winslow, Secretary, Brandon, Vermont, being similar to that adopted in Scotland in 1884, and changed in a few points to render them applicable to this country.

POINTS.	SCALE OF POINTS FOR BULL.	COUNTS.
1.	Head and Horns,	10
2.	Neck,	10
3.	Forequarters,	7
4.	Back and Ribs,	10
5.	Hindquarters,	10
6.	Scrotum,	7
7.	Legs,	5
8.	Skin,	10
9.	Color,	8
10.	Weight,	10
11.	General Appearance,	15
12.	Escutcheon,	8
	PERFECTION,	100

The points desirable in the female are generally so in the male, but must, of course, be attended with that masculine character which is inseparable from a strong and vigorous constitution. Even a certain degree of coarseness is admissible; but then it must be so exclusively of masculine description as never to be discovered in a female of his get.

DETAILED DESCRIPTION.

POINTS. COUNTS.

1. HEAD.—Head of the bull may be shorter than that of the cow, but the frontal bone should be broad, the muzzle of good size, throat nearly free from hanging folds; eyes full. The horns should have an upward turn, with sufficient size at the base to indicate strength of constitution, 10

2. NECK.—Of medium length, somewhat arched, and large in those muscles which indicate power and strength, . 10

3. FOREQUARTERS.—Shoulders close to the body, without any hollow space behind; chest broad, brisket deep and well

developed, but not too large, ... 7
4. BACK.—Short and straight; spine sufficiently defined, but not in the same degree as in the cow; ribs well sprung, and body deep in the flanks, ... 10
5. HINDQUARTERS.—Long, broad and straight; hip bones wide apart; pelvis long, broad and straight; tail set on a level with the back; thighs deep and broad, . . . 10
6. SCROTUM.—Large with well developed teats in front, . . 7
7. LEGS.—Short in proportion to size, joints firm. Hind legs well apart, and not to cross in walking, ... 5
8. SKIN.—Yellow, soft, elastic, and of medium thickness, . 10
9. COLOR.—Red of any shade, brown or white, or a mixture of these—each color being distinctly defined, . . 3
10. WEIGHT.—Average live weight at maturity, about 1500, . 10
11. GENERAL APPEARANCE—including style and movement, . 15
12. ESCUTCHEON.—Large and fine development, ... 3
      PERFECTION, - - - - 100

POINTS.	SCALE OF POINTS FOR COW.	COUNTS.
1.	Head and Horns,	10
2.	Neck,	5
3.	Forequarters,	5
4.	Back and Ribs,	10
5.	Hindquarters,	8
6.	Udder, Milk Veins and Teats,	30
7.	Legs and Bones,	8
8.	Skin and Hair,	5
9.	Color,	3
10.	Weight,	8
11.	General Appearance,	10
12.	Escutcheon,	8
	PERFECTION,	100

DETAILED DESCRIPTION.

POINTS.   COUNTS.

1. HEAD.—Short; forehead, wide; nose, fine between the muzzle and eyes; muzzle large; eyes full and lively; horns wide set on, inclining upwards, ... 10
2. NECK.—Moderately long, and straight from the head to the top of the shoulder, free from loose skin on the underside, fine at its junction with the head, and enlarging symmet-

rically towards the shoulders, ... 5
3. FOREQUARTERS.—Shoulders, sloping; withers, fine; chest, sufficiently broad and deep to insure constitution; brisket and whole forequarters light, the cow gradually increasing in depth and width backwards, ... 5
4. BACK.—Short and straight; spine, well defined, especially at the shoulders; short ribs, arched; the body deep at the flanks, ... 10
5. HINDQUARTERS.—Long, broad and straight, hookbones wide apart, and not overlaid with fat; thighs deep and broad; tail, long, slender, and set on level with the back, . . 8
6. UDDER.—Capacious, and not fleshy, hindpart broad and firmly attached to the body, the sole nearly level and extending well forward; milk veins about udder and abdomen well developed; the teats from 2½ to 3 inches in length, equal in thickness—the thickness being in proportion to the length—hanging perpendicularly; their distance apart *at the sides* should be equal to one-third of the length of the vessel, and *across* to about one-half of the breadth, ... 30
7. LEGS.—Short in proportion to size, the bones fine, the joints firm, ... 8
8. SKIN.—Yellow, soft and elastic, and covered with soft, close, woolly hair, ... 5
9. COLOR.—Red of any shade, brown or white, or a mixture of these—each color being distinctly defined, . . . 3
10. WEIGHT.—Average live weight, in full milk, about 1,000 pounds, ... 8
11. GENERAL APPEARANCE, including style and movement, . 10
12. ESCUTCHEON.—Large and fine development. . . . 8

PERFECTION, - - - - 100

DEVON CATTLE.

Standard of Excellence for Devon Cattle as adopted by the American Devon Cattle Club, L. P. Sisson, Secretary, Wheeling, West Virginia.

POINTS.	SCALE OF POINTS FOR BULL.	COUNTS.
1.	Head and Horns,	10
2.	Cheek,	2
3.	Neck,	4
4.	Shoulders,	6
5.	Chest,	10
6.	Ribs,	10
7.	Back, Loin and Rump,	20
8.	Hindquarters,	12
9.	Tail and Switch,	2
10.	Legs,	4
11.	Skin, Color and Hair,	8
12.	Size,	4
13.	General Appearance,	8
	PERFECTION,	100

DETAILED DESCRIPTION.

POINTS. COUNTS.

1. HEAD.—Masculine, full and broad, tapering toward the nose, which should be flesh-colored; nostrils high and open, muzzle broad; eye full and placid and surrounded with flesh-colored ring; ears of medium size and thickness; horns medium size, growing at right angles from the head, or slightly elevated, waxy at the base, tipped with a darker shade, ... 10

2. CHEEK.—Full and broad at root of tongue; throat clean, . 2

3. NECK.—Of medium length and muscular, widening from the head to the shoulders, and strongly set on, . . . 4

4. SHOULDERS.—Fine, flat, sloping and well fleshed, arms strong with firm joints, ... 6

5. CHEST.—Deep, broad and somewhat circular, . . . 10

6. RIBS.—Well sprung from the back-bone, nicely arched, deep, with flanks fully developed, ... 10

7. BACK.—Straight and level from the withers to the setting on of the tail; loin broad and full; hips and rump of medium width and on a level with the back, ... 20
8. HINDQUARTERS.—Deep, thick and square, ... 12
9. TAIL.—Well set on at a right angle with the back, tapering, with a switch of white or roan hair and reaching the hocks, ... 2
10. LEGS.—Short, straight and squarely placed when viewed from behind, not to cross or sweep in walking, hoof well formed, ... 4
11. SKIN.—Moderately thick and mellow, covered with an abundant coat of rich hair of a red color; no white spot admissible unless around the purse, ... 8
12. SIZE.—Minimum weight at three years old 1,400 pounds, . 4
13. GENERAL APPEARANCE.—As indicated by stylish and quick movement, form, constitution, and vigor, and the underline as nearly as possible parallel with the line of the back, ... 8

PERFECTION, - - - - 100

Purity of blood must be evidenced by registry in the American Devon Record.

POINTS. SCALE OF POINTS FOR COW. COUNTS.
1. Head and Horns, ... 8
2. Neck, ... 4
3. Shoulders, ... 4
4. Chest, ... 8
5. Ribs, ... 8
6. Back, Loin and Rump, ... 16
7. Hindquarters, ... 8
8. Udder and Teats, ... 20
9. Tail and Switch, ... 2
10. Legs, ... 4
11. Skin, Color and Hair, ... 8
12. Size, ... 2
13. General Appearance, ... 8

PERFECTION, - - - 100

DETAILED DESCRIPTION.
POINTS. COUNTS.
1. HEAD.—Moderately long, with a broad, indented forehead, tapering considerably towards the nostrils; the nose of a

flesh color, nostrils high and open, the jaws clean, the eye bright, lively and prominent, and surrounded by a flesh colored ring, throat clean, ears thin, the expression gentle and intelligent; horns matching, spreading and gracefully turned up, of waxey color, tipped with a darker shade, ... 8

2. NECK.—Upper line short fine at head, widening and deep at withers and strongly set to the shoulders, ... 4
3. SHOULDERS.—Fine, flat and sloping, with strong arms and firm joints, ... 4
4. CHEST —Deep, Broad, and somewhat circular in character, 8
5. RIBS.—Well sprung from the back-bone, nicely arched, deep, with flanks fully developed, ... 8
6. BACK.—Straight and level from the withers to the setting on of the tail; loin broad and full; hips and rump of medium width, and on a level with the back, . . 16
7. HINDQUARTERS.—Deep, thick and square, ... 8
8. UDDER.—Not fleshy, coming well forward in line with the belly and well up behind; teats moderately large, and squarely placed, ... 20
9. TAIL.—Well set on at a right angle with the back, tapering, with a switch of white or roan hair and reaching the hocks, ... 2
10. LEGS.—Straight, squarely placed when viewed from behind, not to cross or sweep in walking; hoof well formed, . 4
11. SKIN.—Moderately thick and mellow, covered with an abundant coat of rich hair of a red color; no white spot admissible, except the udder, ... 8
12. SIZE.—Minimum weight at three years old, 1,000 pounds, . 2
13. GENERAL APPEARANCE.—As indicated by stylish and quick movement, form, constitution and vigor, and the under line as nearly as possible parallel with the line of the back, 8

PERFECTION, - - - - 100

Purity of blood must be evidenced by registry in the American Devon Record.

DUTCH-BELTED CATTLE.

Standard of Excellence for Dutch-Belted Cattle, as adopted by the Dutch-Belted Cattle Association of America, H. B. Richards, Secretary, Easton, Pa.

SCALE OF POINTS FOR BULL.

POINTS.		COUNTS.
1.	Body-color and Belt,	18
2.	Head, Muzzle and Tongue,	6
3.	Eyes and Horns,	4
4.	Neck,	6
5.	Shoulders,	9
6.	Barrel and Ribs,	10
7.	Hips, Chine and Loin,	10
8.	Rump,	6
9.	Hindquarters, Tail and Switch,	8
10.	Legs,	8
11.		
12.	Escutcheon,	2
13.	Hair and Skin,	3
14.	Disposition,	4
15.	General Condition,	6
16.	Rudimentary Teats,	10
	PERFECTION,	100

The scale of points for males shall be the same as those given for females, except that No. 11 shall be omitted, and the bull credited 10 points for size and wide-spread placing of rudimentary teats, 5 points additional for development of shoulder, and 5 additional points for perfection of belt.

DETAILED DESCRIPTION.

POINTS. COUNTS.

1. BODY-COLOR.—Black, with a clearly defined continuous white belt. The belt to be of medium width, beginning behind the shoulder and extending nearly to the hips, 18

2. HEAD.—Comparatively long and somewhat dishing; broad between the eyes. Poll, prominent; muzzle fine; dark tongue, ... 6

3. EYES.—Black, full and mild. Horns, long compared with
their diameter, ... 4
4. NECK.—Fine and moderately thin, and should harmonize in
symmetry with the head and shoulders, . . . 6
5. SHOULDERS.—Fine at the top, becoming deep and broad as
they extend backward and downward, with a low chest, 9
6. BARREL.—Large and deep, with well-developed abdomen;
ribs well rounded and free from fat, ... 10
7. HIPS.—Broad and chine, level, with full loin, . . 10
8. RUMP.—High, long and broad, ... 6
9. HINDQUARTERS.—Long and deep, rear line incurving. Tail
long, slim, tapering to a full switch, . . . 8
10. LEGS.—Short, clean, standing well apart, . . . 3
11. . . . .
12. ESCUTCHEON.— 2
13. HAIR.—Fine and soft; skin of moderate thickness, of a rich
dark or yellow color, ... 3
14. DISPOSITION.—Quiet, and free from excessive fat, . . . 4
15. GENERAL CONDITION and apparent constitution, . . 6
16. RUDIMENTARY TEATS.—For size and wide spread placing of
rudimentary teats, ... 10
———
PERFECTION, - - - - 100

POINTS.	SCALE OF POINTS FOR COW.	COUNTS.
1.	Body-color and Belt,	8
2.	Head, Muzzle and Tongue,	6
3.	Eyes and Horns,	4
4.	Neck,	6
5.	Shoulders,	4
6.	Barrel and Ribs,	10
7.	Hips, Chine and Loin,	10
8.	Rump,	6
9.	Hindquarters, Tail and Switch,	8
10.	Legs,	3
11.	Udder, Teats and Mammary Veins,	20
12.	Escutcheon,	2
13.	Hair and Skin,	3
14.	Disposition,	4
15.	General Condition,	6
	PERFECTION,	100

DETAILED DESCRIPTION.

POINTS.   COUNTS.

1. BODY-COLOR.—Black, with a clearly-defined continuous white belt. The belt to be of medium width, beginning behind the shoulder and extending nearly to the hips, . . . 8
2. HEAD.—Comparatively long and somewhat dishing; broad between the eyes. Poll, prominent; muzzle, fine; dark tongue, ... 6
3. EYES.—Black, full and mild. Horns, long compared with their diameter, ... 4
4. NECK.—Fine and moderately thin and should harmonize in symmetry with the head and shoulders, ... 6
5. SHOULDERS.—Fine at the top, becoming deep and broad as they extend backward and downward, with a low chest, 4
6. BARREL.—Large and deep, with well-developed abdomen; ribs well rounded and free from fat, ... 10
7. HIPS.—Broad and chine, level, with full loin, . . . 10
8. RUMP.—High, long and broad, ... 6
9. HINDQUARTERS.—Long and deep, rear line incurving. Tail long, slim, tapering to a full switch, ... 8
10. LEGS.—Short, clean, standing well apart, . . . 8
11. UDDER.—Large, well-developed front and rear. Teats of convenient size and wide apart; mammary veins large, long and crooked, entering large orifices, ... 20
12. ESCUTCHEON, ... 2
13. HAIR.—Fine and soft; skin of moderate thickness of a rich dark or yellow color, ... 3
14. DISPOSITION.—Quiet and free from excessive fat, . . . 4
15 GENERAL CONDITION and apparent constitution, . . . 6

                          PERFECTION, - - - - 100

GALLOWAY CATTLE.

Standard of Excellence for Galloway Cattle, compiled by L. P. Muir, Secretary of the American Galloway Breeders' Association, Independence, Mo., from a detailed description drawn up in 1883 by the Council of the Galloway Society of Great Britain. [This standard was kindly sent to me by Mr. Muir at my request, for this publication, and only to be used until the American Galloway Breeders' Association shall adopt one of their own.—ED.]

SCALE OF POINTS FOR GALLOWAY CATTLE.

POINTS.	COUNTS.
1. Color.	8
2. Head,	5
3. Eye,	2
4. Ear,	2
5. Neck,	8
6. Body,	10
7. Shoulders,	6
8. Breast,	8
9. Back and Rump,	8
10. Ribs,	8
11. Loin and Sirloin,	10
12. Hook Bones,	2
13. Hindquarters,	8
14. Flank,	4
15. Thighs,	4
16. Legs,	4
17. Tail,	3
18. Skin,	5
19. Hair,	5
PERFECTION,	100

DETAILED DESCRIPTION.

POINTS. COUNTS.

1. COLOR.—Black, with a brownish tinge, ... 3
2. HEAD.—Short and wide, with broad forehead and wide nostrils, without the slightest symptoms of horns or scurs, . 5
3. EYE.—Large and prominent, ... 2
4. EAR.—Moderate in length, and broad, pointing forward and upward with fringe of long hair, ... 2

GALLOWAY CATTLE.

5. NECK.—Medium in length; clean and fitting well into the shoulders, the top in a line with the back in a female, and in a male naturally rising with age, ... 8
6. BODY.—Rounded, deep and symmetrical, ... 10
7. SHOULDERS.—Fine and straight; moderately wide above. Coarse shoulder points, and sharp or high shoulders are objectionable, ... 6
8. BREAST.—Full and deep, ... 8
9. BACK AND RUMP.—Straight, ... 8
10. RIBS.—Deep and well sprung, ... 8
11. LOIN AND SIRLOIN.—Well filled, ... 10
12. HOOK BONES.—Not prominent, ... 2
13. HINDQUARTERS.—Long, moderately wide and well filled, . 8
14. FLANK.—Deep and full, ... 4
15. THIGHS.—Broad, straight and well let down to hock. Rounded buttocks are very objectionable, ... 4
16. LEGS.—Short and clean with fine bone, ... 4
17. TAIL.—Well set on and moderately thick, ... 3
18. SKIN.—Mellow and moderately thick, ... 5
19. HAIR.—Soft and wavy, with mossy undercoat. Wiry or curly coarse hair is very objectionable, ... 5

        PERFECTION, - - - - - 100

GUERNSEY CATTLE.

Standard of Excellence for Guernsey Cattle, as adopted by the American Guernsey Cattle Club, W. H. Caldwell, Secretary, Peterboro, N. H., also by the Guernsey Breeders' Association, W. B. Harvey, Secretary, West Grove, Pa.—[Slightly changed in arrangement for this publication.]

SCALE OF POINTS FOR BULL.

POINTS.	COUNTS.
1. Color of Skin,	20
2. Handling of Skin and Hair,	10
3. Escutcheon,	8
4. Milk Veins,	6
5, 6 and 7,	
8. Position of Teats,	4
9. Size of Teats,	4
10. Size,	5
11. Bone,	1
12. Barrel,	4
13. Hips and Loin,	2
14. Rump,	2
15. Thighs and Withers,	2
16. Back,	3
17. Throat,	1
18. Legs,	2
19. Tail,	1
20. Horns,	2
21. Head,	3
22. General Appearance,	2
PERFECTION,	82

For Bulls deduct 18 counts for udder, points 5, 6 and 7.

DETAILED DESCRIPTION.

POINTS.	COUNTS.
1. SKIN.—Deep yellow in ear, on end of bone of tail, at base of horns, on udder, teats and body generally,	20
2. SKIN.—Loose, mellow, with fine, soft hair,	10
3. ESCUTCHEON.—Wide on thighs, high and broad, with thigh ovals,	8
4. MILK VEINS.—Long and prominent,	6
5, 6 and 7,	
8. UDDER TEATS.—Squarely placed,	4

GUERNSEY CATTLE.

9. UDDER TEATS.—Of good size, ... 4
10. SIZE.—For the breed, ... 5
11. BONE.—Not too light, ... 1
12. BARREL.—Round and deep at flank, ... 4
13. HIPS AND LOIN.—Wide, ... 2
14. RUMP.—Long and broad, ... 2
15. THIGHS AND WITHERS.—Thin, ... 2
16. BACK.—Level to setting on of tail, ... 3
17. THROAT.—Clean, with small dewlap, ... 1
18. LEGS.—Not too long, with hocks well apart in walking, . 2
19. TAIL.—Long and thin, ... 1
20. HORNS.—Curved and not coarse, ... 2
21. HEAD.—Rather long and fine, with quiet and gentle expression, ... 3
22. GENERAL APPEARANCE, ... 2

PERFECTION, - - - - 82

For Bulls deduct 18 counts for udder, points 5, 6 and 7.

POINTS.	SCALE OF POINTS FOR COW.	COUNTS.
1.	Color of skin,	20
2.	Handling of skin and hair,	10
3.	Escutcheon,	8
4.	Milk veins,	6
5.	Udder in front,	6
6.	Udder behind,	8
7.	Size of udder,	4
8.	Position of teats,	4
9.	Size of teats,	4
10.	Size,	5
11.	Bone,	1
12.	Barrel,	4
13.	Hips and loin,	2
14.	Rump,	2
15.	Thighs and withers,	2
16.	Back,	3
17.	Throat,	1
18.	Legs,	2
19.	Tail,	1
20.	Horns,	2
21.	Head,	3
22.	General appearance,	2

PERFECTION, - - - 100

DETAILED DESCRIPTION.

POINTS.		COUNTS
1.	SKIN.—Deep yellow, in ear, on end of bone of tail, at base of horns, on udder, teats and body generally,	20
2.	SKIN.—Loose, mellow, with fine, soft hair,	10
3.	ESCUTCHEON.—Wide on the thighs, high and broad, with thigh ovals,	8
4.	MILK VEINS.—Long and prominent,	6
5.	UDDER.—Full in front,	6
6.	UDDER.—Full and well up behind,	8
7.	UDDER.—Large, but not fleshy,	4
8.	UDDER TEATS.—Squarely placed,	4
9.	UDDER TEATS.—Of good size,	4
10.	SIZE—For the breed,	5
11.	BONE.—Not too light,	1
12.	BARREL.—Round and deep at flank,	4
13.	HIPS AND LOIN.—Wide,	2
14.	RUMP.—Long and broad,	2
15.	THIGHS AND WITHERS.—Thin,	2
16.	BACK.—Level to setting on of tail,	3
17.	THROAT.—Clean, with small dewlap,	1
18.	LEGS.—Not too long, with hocks well apart in walking,	2
19.	TAIL.—Long and thin,	1
20.	HORNS.—Curved and not coarse,	2
21.	HEAD.—Rather long and fine, with quiet and gentle expression	3
22.	GENERAL APPEARANCE,	2
	PERFECTION,	100

HEREFORD CATTLE.

In answer to an enquiry from the editor of this publication, Mr. C. R. Thomas, of Independence, Mo., the secretary of the American Hereford Cattle Breeders' Association, wrote that the association had not adopted an official standard of excellence, but advised the editor to apply for one to Mr. S. W. Anderson, of Asbury, W. Va., who is quite a large breeder, and is considered a good Hereford judge. Acting upon the suggestion of the Secretary, the editor corresponded with Mr. Anderson, who thereupon kindly furnished the following standard of excellence for Herefords.

POINTS.	SCALE OF POINTS FOR HEREFORD CATTLE.	COUNTS.
1.	Color,	3
2.	Head,	10
3.	Horns,	4
4.	Neck,	4
5.	Shoulders,	6
6.	Heart-girth,	10
7.	Chest,	6
8.	Brisket,	4
9.	Ribs,	8
10.	Back,	13
11.	Hindquarters,	8
12.	Tail,	2
13.	Underline,	4
14.	Legs,	4
15.	Flesh,	5
16.	Size,	4
17.	Skin,	6
	Perfection,	100

DETAILED DESCRIPTION.

POINTS.   COUNTS.

1. COLOR.—In color the Hereford should invariably be a red (either light or dark), with white face, throat, chest, lower part of the body and legs, together with the crest or mane, tip of tail; and, generally a white strip along the withers,   3

2. HEAD.—The bull should have a good masculine head; broad between the eyes, which should be full and lively. The countenance should present a placid appearance, denoting a good temper and that quietude of disposition so essential to the successful grazing of all ruminating animals. The cow's head should be much the same, but finer. The nose should be a pure white or flesh color. The cheeks

and throat should be full, with tongue root large and loose, 10
3. HORNS.—The horns of the bull should be large at the butt, and a good length is not objectionable. Those of the cow should be long, but much smaller. The horns of either should be waxey white, although they are occasionally found dark at the points, ... 4
4. NECK.—Short and meaty, and well set on the shoulders, . 4
5. SHOULDERS.—Deep, sloping, thick and fleshy. So beautifully should the shoulder blades blend into the body that it would be difficult to tell in a well-fed animal where they are set on, ... 6
6. HEART-GIRTH.—Full and deep in foreflank behind the arm; full behind the shoulders without depression, and broad across the crops, ... 10
7. CHEST.—Expanded, deep and full; well covered on the outside with mellow flesh, ... 6
8. BRISKET.—Well developed, and projecting firmly from between the legs, proportionately covered with flesh and fat, 4
9. RIBS.—Well sprung, wide, and evenly covered with flesh, . 8
10. BACK.—Straight and level from crops to hips, which latter should be moderately broad; loin, strong, wide and deep, 12
11. HINDQUARTERS.—Should be long from the hip back; the rump forming a straight line with the back, and at a right angle with the thigh, which should be full of flesh down to the hocks; twist good, well filled up with flesh even with the thigh, ... 8
12. TAIL.—Well set on and falling in a plumb line to the hocks, 2
13. UNDERLINE.—As nearly straight as possible; the flank full and about on a straight line with the belly, . . . 4
14. LEGS.—Short and well apart; muscular hocks and knees, . 4
15. FLESH.—The whole carcass well and evenly covered with a rich, mellow flesh, ... 5
16. SIZE.—Minimum weight for bull at three years old, 1,800 pounds; minimum weight for cow at three years old, 1,500 pounds, ... 4
17. SKIN.—The hide, thick, yet mellow, and well covered with soft, glossy hair, having a tendency to curl; the hide giving the impression when you touch it that it will stretch to any extent, ... 6

PERFECTION, - - - 100

THE AMERICAN HOLDERNESS CATTLE.

The Holderness is a pure-breed of cattle raised for many years by T. A. Cole, of Solsville, N. Y., and numbering several hundred head, of excellent animals. Whilst no standard of excellence has yet been adopted for them, the following description by Lewis F. Allen, author of "American Cattle," will be of interest:

"Mr. Cole calls his cattle "Holderness," a name seldom heard of at the present day; yet several specimens of such a breed of cattle were imported 50 years ago into Massachusetts, to my certain knowledge, and I believe also into this State, and Mr. Knox, from whom Mr. Cole bought his original cow, stated that she was from imported stock, but who was the importer or how long ago the importation was made, was not ascertained. In Youatt's "Cattle, their Breeds and Management," published in London in the year 1834, in which the several breeds of England are described, is mentioned the "Holderness," existing in a district of that name, in the West Riding of Yorkshire. They are noted as great milkers, a branch probably of the ancient unimproved Shorthorns, as they resemble that breed more than any other, except in color and rotundity of form.

The imported ones of fifty years ago were usually dark red or brown on the sides, striped with white on the backs and bellies, and occasionally a little spotted, and Mr. Cole's cows much resemble them in form and size. So there can be little if any doubt of the descent of the latter from the original importations. Yet a marked change in color has been developed in Mr. Cole's herd. His original cow was light red on her sides, with white back and belly. Her first bull calf and several succeeding ones of the herd were also of these colors. But gradually they began in calfhood, and in successive years, to turn the red into black, and now the color of nearly every one of the grown cows and bulls is a dark brown or jet black and white line-back. Why the colors are so changed is a physiological secret. The fact is positive, and their intense in-and-in breeding may have set them back to the color of their long ago ancestry in Holderness."

HOLSTEIN-FRIESIAN CATTLE.

Standard of Excellence for Holstein-Friesian Cattle, as adopted by the Holstein-Friesian Association of America, F. L. Houghton, Secretary, Brattleboro, Vermont; also by the American Branch Association of the North Holland Herd Book, F. H. Beach, Secretary, No. 6 Harrison street, New York; also by the Holstein-Friesian Association of Canada, G. W. Clemens, Secretary, St. George, Ont.

SCALE OF POINTS FOR BULL.

POINTS.	COUNTS.
1. Head,	2
2. Forehead,	2
3. Face,	2
4. Ears,	1
5. Eyes,	2
6. Horns,	2
7. Neck.	5
8. Shoulders,	4
9. Chest,	8
10. Crops,	4
11. Chine,	3
12. Barrel,	6
13. Loin and Hips,	5
14. Rump,	5
15. Throat,	4
16. Quarters,	5
17. Flanks,	2
18. Legs and Feet,	6
19. Tail and Switch,	2
20. Hair and Handling,	10
21. Mammary Veins,	10
22. Rudimentary Teats,	2
23. Escutcheon,	8
Perfection,	100

DETAILED DESCRIPTION.

POINTS.	COUNTS.
1. HEAD.—Showing full vigor; elegant in contour,	2
2. FOREHEAD.—Broad between the eyes; dishing,	2
3. FACE.—Of medium length; clean and trim, especially under the eyes; the bridge of the nose straight; the muzzle broad,	2

4. EARS.—Of medium size, of fine texture; the hair plentiful and soft; the secretions oily and abundant, . . . 1
5. EYES.—Large, full, mild, bright, ... 2
6. HORNS.—Short, of medium size at base, gradually diminishing toward tips; oval, inclining forward; moderately curved inward; of fine texture; in appearance, waxy, 2
7. NECK.—Long, finely crested (if the animal is mature); fine and clean at juncture with the head; nearly free from dewlap; strongly and smoothly joined to shoulders, . 5
8. SHOULDERS.—Of medium height, of medium thickness and smoothly rounded at tops; broad and full at sides; smooth over front, ... 4
9. CHEST.—Deep and low; well filled and smooth in the brisket; broad between the forearms; full in the foreflanks [or through at the heart], ... 8
10. CROPS.—Comparatively full, nearly level with the shoulders, 4
11. CHINE.—Straight; broadly developed; open, . . 8
12. BARREL.—Well rounded, with large abdomen; strongly and trimly held up, ... 6
13. LOIN AND HIPS.—Broad, level or nearly level between hookbones; level and strong laterally; spreading from the chine broadly and nearly level; the hook-bones fairly prominent, ... 5
14. RUMP.—Long, broad, high; nearly level laterally; comparatively full above the thurl, ... 5
15. THURL.—High; broad, ... 4
16. QUARTERS.—Deep, broad; straight behind; wide and full at sides; open and well arched in the twist, . . . 5
17. FLANKS.—Deep; full, ... 2
18. LEGS AND FEET.—Comparatively short, clean and nearly straight; wide apart; firmly and squarely set under the body; arms wide, strong and tapering; feet of medium size, round, solid and deep, ... 6
19. TAIL AND SWITCH.—Large at base, the setting well back; tapering finely to switch; the end of the bone reaching to hocks or below; the switch full, ... 2
20. HAIR AND HANDLING.—Hair healthful in appearance; fine, soft and furry; skin of medium thickness and loose; mellow under the hand; the secretions oily, abundant and of a rich brown or yellow color, ... 10

21. **MAMMARY VEINS.**—Large; full; entering large or numerous orifices; double extension; with special developments, such as forks, branches, connections, etc., ... 10
22. **RUDIMENTARY TEATS.**—Large, well placed, ... 2
23. **ESCUTCHEON.**—Largest; finest, ... 8

      PERFECTION, - - - - 100

POINTS.	SCALE OF POINTS FOR COW.	COUNTS.
1.	Head,	2
2.	Forehead,	2
3.	Face,	2
4.	Ears,	1
5.	Eyes,	2
6.	Horns,	2
7.	Neck,	4
8.	Shoulders,	3
9.	Chest,	6
10.	Crops,	2
11.	Chine,	3
12.	Barrel,	4
13.	Loin and Hips,	5
14.	Rump,	5
15.	Thurl,	4
16.	Quarters,	4
17.	Flanks,	2
18.	Legs,	5
19.	Tail,	2
20.	Hair and Handling,	10
21.	Mammary Veins,	10
22.	Udder and Teats,	12
23.	Escutcheon,	8

      PERFECTION, - - - - 100

DETAILED DESCRIPTION.

POINTS.   COUNTS.

1. **HEAD.**—Decidedly feminine in appearance; fine in contour, 2
2. **FOREHEAD.**—Broad between the eyes; dishing, . . . 2
3. **FACE.**—Of medium length; clean and trim, especially under the eyes, showing facial veins; the bridge of the nose straight; the muzzle broad, ... 2
4. **EARS.**—Of medium size, of fine texture; the hair plentiful

HOLSTEIN-FRIESIAN CATTLE.

and soft; the secretions oily and abundant, . . . 1
5. EYES.—Large, full, mild and bright, ... 2
6. HORNS.—Small, tapering finely toward the tips; set moderately narrow at base; oval; inclining forward; well bent inward; of fine texture; in appearance, waxy, 2
7. NECK.—Long; fine and clean at juncture with the head; free from dewlap; evenly and smoothly joined to shoulders, ... 4
8. SHOULDERS.—Slightly lower than hips; fine and even over tops; moderately broad and full at sides, ... 8
9. CHEST.—Of moderate depth and lowness; smooth and moderately full in the brisket; full in the foreflanks [or through at the heart], ... 6
10. CROPS.—Moderately full, ... 2
11. CHINE.—Straight; broadly developed; open, . . 8
12. BARREL.—Of wedge shape; well rounded; with a large abdomen; trimly held up [in judging the last item, age must be considered], ... 4
13. LOINS AND HIPS.—Broad; level, or nearly level between the hook bones; level and strong laterally; spreading from chine broadly and nearly level; hook bones fairly prominent, ... 5
14. RUMP.—Long, high; broad, with roomy pelvis; nearly level laterally; comparatively full above the thurl, . . 5
15. THURL.—High; broad, ... 4
16. QUARTERS.—Deep; straight behind; roomy in the twist; wide and moderately full at the sides, ... 4
17. FLANKS.—Deep; comparatively full, ... 2
18. LEGS.—Comparatively short; clean and nearly straight; wide apart; firmly and squarely set under the body; feet of medium size, round, solid and deep, ... 5
19. TAIL.—Large at base, the setting well back; tapering finely to switch; the end of the bone reaching to the hocks or below; the switch full, ... 2
20. HAIR AND HANDLING.—Hair healthful in appearance; fine, soft and furry; the skin of medium thickness and loose; mellow under the hand; the secretions oily, abundant and of a rich brown or yellow color, ... 10
21. MAMMARY VEINS.—Very large; very crooked [age must be taken into consideration in judging of size and crooked-

ness]; entering very large or numerous orifices; double extension; with special developments, such as branches, connections, etc., ... 10
22. UDDER AND TEATS.—Very capacious; very flexible; quarters even, nearly filling the space in the rear below the twist, and extending well forward in front; broad and well held up; teats well formed, wide apart, plumb and of convenient size, ... 12
23. ESCUTCHEON.—Largest; finest, ... 8

                PERFECTION, - - - - 100

JERSEY CATTLE.

Standard of Excellence for Jersey cattle, as adopted May 6th, 1885, by the American Jersey Cattle Club, J. J. Hemingway, secretary, 8 West 17th street, New York.

SCALE OF POINTS FOR BULL.

POINTS.		COUNTS.
1. Head and Face,		2
2. Eyes and Horns,		1
3. Neck,		8
4. Back,		1
5. Loins,		6
6. Barrel,		10
7. Hip and Rump,		10
8. Legs,		2
9. Tail and Switch,		1
10. Color and Handling,		5
11 and 12,		
13. Teats,		10
14.		
15. Disposition,		5
16. General Appearance,		10
PERFECTION,		71

The same scale of points as for cows shall be used in judging bulls, omitting Nos. 11, 12 and 14, and making due allowance for masculinity; but when bulls are exhibited with their progeny, in a separate class, add 30 counts for their progeny.

DETAILED DESCRIPTION.

POINTS.	COUNTS.

1. HEAD.—Small and lean; face dished, broad between the eyes and narrow between the horns, ... 2
2. EYES.—Full and placid; horns small, crumpled, and amber-colored, ... 1
3. NECK.—Thin, rather long, with clean throat, and not heavy at the shoulders, ... 8
4. BACK.—Level to the setting-on of the tail, ... 1
5. LOINS.—Broad across the loins, ... 6

6. BARREL.—Long, hooped, broad, and deep at the flank, . 10
7. HIPS.—Wide apart; rump long, ... 10
8. LEGS.—Short, ... 2
9. TAIL.—Fine, reaching the hocks, with good switch, . . 1
10. COLOR.—Color and mellowness of hide; inside of ears yellow, ... 5
11 and 12. . . . .
13. TEATS.—Rather large, wide apart, and squarely placed, . 10
14. . . . .
15. DISPOSITION.—Quiet, ... 5
16. GENERAL APPEARANCE and apparent constitution, . . 10

                PERFECTION, - - - 71

POINTS.	SCALE OF POINTS FOR COW.	COUNTS.
1.	Head and Face,	2
2.	Eyes and Horns,	1
3.	Neck,	8
4.	Back,	1
5.	Loins,	6
6.	Barrel,	10
7.	Hips and Rump,	10
8.	Legs,	2
9.	Tail and Switch,	1
10.	Color and Handling,	5
11.	Fore Udder,	13
12.	Hind Udder,	11
13.	Teats,	10
14.	Milk Veins,	5
15.	Disposition,	5
16.	General Appearance,	10
	PERFECTION,	100

In judging heifers, omit Nos. 11, 12 and 14.

DETAILED DESCRIPTION.

POINTS.   COUNTS.

1. HEAD.—Small and lean; face dished, broad between the eyes and narrow between the horns, ... 2
2. EYES.—Full and placid; horns small, crumpled and amber-colored, ... 1
3. NECK.—Thin, rather long, with clean throat, and not heavy at shoulders, ... 8

JERSEY CATTLE.

4. BACK.—Level to the setting on of tail, ... 1
5. LOINS.—Broad across the loins, ... 6
6. BARREL.—Long, hooped, broad, and deep at the flank, . 10
7. HIPS.—Wide apart, rump long, ... 10
8. LEGS.—Short, ... 2
9. TAIL.—Fine, reaching the hocks with good switch, . . 1
10. COLOR.—Color and mellowness of hide; inside of ears yellow, ... 5
11. FORE-UDDER.—Full in form and not fleshy. 13
12. HIND-UDDER.—Full in form and well up behind, . . 11
13. TEATS.—Rather large, wide apart, and squarely placed, . 10
14. MILK VEINS.—Prominent, ... 5
15. DISPOSITION.—Quiet, ... 5
16. GENERAL APPEARANCE and apparent constitution, . . 10

                      PERFECTION, - - - - 100

In judging heifers, omit Nos. 11, 12 and 14.

KERRY CATTLE.

In the United States quite a number of Kerry cattle have been imported from time to time, but as there is neither a Kerry Cattle Club, nor a Kerry herd book in this country, their valuable characteristics are not as fully recognized here as they should be. The cows are good milkers for their size (some weighing only 300 pounds), giving from 10 to 20 quarts daily, and, whilst, from the ease with which they are kept in a limited space, they are often called the poor man's cow, they might just as truly be termed the rich man's cow, in that they are small and handsome, and so docile that they can be easily tethered on a lawn. The following Standard of Excellence, approved by Mr. Henry S. Ambler, of Chatham, N. Y., the principal breeder of Kerry Cattle in America, may be of assistance to judges and breeders:

SCALE OF POINTS FOR KERRY CATTLE.

POINTS.		COUNTS.
1. Head,		2
2. Cheeks,		1
3. Throat,		1
4. Muzzle,		2
5. Nostrils,		1
6. Horns,		2
7. Ears,		2
8. Eyes,		1
9. Neck,		6
10. Shoulders,		3
11. Chest,		6
12. Barrel,		6
13. Ribs,		4
14. Back,		2
15. Rump and Thigh,		6
16. Tail,		2
17. Skin and Handling,		10
18. Udder,		12
19. Teats,		10
20. Milk Veins,		8
21. Fore-legs,		2
22. Hind-legs,		2
23. Hoofs,		1
24. Color,		3
25. Disposition,		5
PERFECTION,		100

In judging bulls and heifers, omit Nos. 18 and 20.

DETAILED DESCRIPTION.

POINTS.   COUNTS.

1. HEAD.—Fine and small and tapering, ... 2
2. CHEEKS.—Clean, ... 1
3. THROAT.—Full, and well set, ... 1
4. MUZZLE.—Fine, and of a rich black color, . . . 2
5. NOSTRILS.—Well placed and rather open. . . . 1
6. HORNS.—Well sprung; smooth; rather thick at base, but gently tapering; white in color, with black tips, . . 2
7. EARS.—Small, fine, and of a fine, pink-orange color within, . 2
8. EYES.—Mild and full, but animated. 1
9. NECK.—Straight and fine; evenly and smoothly joined to the shoulders, ... 6
10. SHOULDERS.—The height at the shoulders should not exceed 4 feet, though 3 feet 6 inches is more desirable, . . . 3
11. CHEST.—Deep and broad, ... 6
12. BARREL.—Deep, full and well hooped, ... 6
13. RIBS.—Well sprung, ... 4
14. BACK.—Even and straight from withers to top of hip, . . 2
15. RUMP.—Rather narrow, long, but straight from top of hips to setting on of tail. Thigh, light, ... 6
16. TAIL.—Long and fine, ... 2
17. SKIN.—Of good rich orange color; loose, mellow, and covered with a good coat of soft hair, ... 10
18. UDDER.—Well rounded, full and capacious; in line with belly and well up behind, ... 12
19. TEATS—Well placed; large and rather far apart, . . 10
20. MILK VEINS.—Very prominent, ... 8
21. FORE LEGS.—Short and straight; full above the knee; fine below, ... 2
22. HIND LEGS.—Not too close together and squarely placed; fine bone, ... 2
23. HOOFS.—Small, ... 1
24. COLOR.—Rich black preferable, although there are some very good animals black and white, and occasionally a few of other colors, ... 3
25. DISPOSITION.—Gentle, ... 5

PERFECTION,   - - - - 100

In judging bulls and heifers omit Nos. 18 and 20.

The Dexter variety is distinguished from the pure or true Kerry in having a round plump body, short and rather thick legs; the head is heavier and wanting in that fineness which marks the true Kerry, and the horns are longer, straighter and coarser.

RED POLLED CATTLE.

The following letters explain themselves and need no further introduction,

RED POLLED CATTLE CLUB OF AMERICA.

SECRETARY'S OFFICE,
Dayton, Ohio, Sept. 22, 189 .

FRANK A. LOVELOCK, SALEM, VA.:

DEAR SIR: Your postal of the 20th received. No regular "scale of points" has been adopted for Red Polls. They should be judged, however, both for beef and dairy qualities, as they claim to excel in both lines.

The most desirable points are smooth, level form, much like the Devons, but larger. A deep, rich red color, without white, except the switch, and it may be a little white; a white spot on the udder. No horns or scurs; a fine head, clean throat, deep body; level rump, broad back or well rounded, and the points generally of a good beef animal.

Then for cows, a large udder and good milk veins, but not always prominent to the eye, on account of the thick, mossy coat of hair. The indications of a good milker, with smooth, compact form, fine bone, rich color, and good size are my own preference.

Very truly,
J. McLAIN SMITH,
Secretary.

RED POLLED CATTLE CLUB OF AMERICA.

SECRETARY'S OFFICE,
Dayton, Ohio, July 4, 1892.

FRANK A. LOVELOCK, SALEM, VA.:

DEAR SIR—Yours of the 2nd received. The Club has not yet adopted any "scale of points," and no movement has been made to that end. There are no changes to make in my former letter. I do not now remember just what I wrote, but I should make very emphatic the requirements of a good show for MILK. Other things being nearly equal, I should give decided preference to the cow showing best milking qualities. Where there is any claim to beef qualities, the strong tendency is to give beef the preference in a show ring. It ought not to be so with Red Polls.

Very truly,
J. McLAIN SMITH,
Secretary.

SHORT-HORN CATTLE.

Mr. J. H. Pickrell, of 510 East Monroe street, Springfield, Illinois, Secretary of the American Short-Horn Breeders' Association, writes the editor of this publication that his association deals primarily with pedigrees and only incidentally with the animals, and has never made up a scale of points. The following standard of excellence, was compiled especially for this work by Col. A. M. Bowman, of Salem, Va., formerly of the firm of Palmer & Bowman (for many years owners of the largest herd of registered Short-Horns in the world), and an ex-member of the Board of Directors of the American Short horn Breeders' Association.

SCALE OF POINTS FOR SHORT-HORN CATTLE.

POINTS.	COUNTS.
1. Head and Face,	8
2. Horns,	3
3. Neck,	3
4. Heart-Girth,	8
5. Shoulders,	6
6. Chest.	8
7. Brisket,	4
8. Crops,	6
9. Ribs,	8
10. Back, Loins and Rump,	12
11. Hindquarters, Thighs and Twist,	10
12. Tail,	2
13. Underline and Flank,	4
14. Legs,	4
15. Flesh,	4
16. Skin, Handling and Hair,	10
PERFECTION,	100

DETAILED DESCRIPTION.

POINTS. COUNTS.

1. HEAD.—Short; forehead broad, gracefully narrowing along the face toward the muzzle; face slightly concaved, but not dished; eye prominent, but with mild expression. Fine, wide, open nostrils; color of nose, yellow or nutty drab; ear should be upright, large, and not too thick, but well covered with a mossy coat of hair, ... 8

2. HORNS.—For bull, strong, but not coarse, standing wide at the base, bending gracefully forward in an outward

curve, and then may incline downward or upward, with waxy or creamy tint, rather than white, and no dark or black except at the tips. For the cow the same general character should prevail, except that the horn should be smaller and finer, ... 8

3. NECK.—Short, well set in the shoulders and tapering to the head, running back on a level in the cow, and with a gradually rising crest in the bull; free from dewlap or hanging skin, ... 3

4. HEART-GIRTH.—Full and deep in fore-flank; full behind the shoulders, without depression, ... 8

5. SHOULDERS.—Broad and even at top, working backward into a level with the chine; smooth at forward points and tapering gracefully to the knees, ... 6

6. CHEST.—Deep, broad and full, without coarseness; a deep, broad and full chest with some coarseness, is however, prefered to a narrow chest, however smooth, . . 8

7. BRISKET.—Prominent, well set forward, almost perpendicular in front; broad and well let down, ... 4

8. CROPS.—Broad and full without depression, ... 6

9. RIBS.—Fore-ribs springing in a well rounded arch from the back-bone, long and deep; hind-ribs should spring well out from the back-bone, long, deep and well set back towards the hips, ... 8

10. BACK.—Spine straight from chine to root of tail; loins broad, full and level with the spine and hips; hips wide-spread, smooth and on a level with the spine; rump long, full broad and level, ... 12

11. HINDQUARTERS.—Should drop perpendicularly from the points of the rump; thighs broad and full and running well down to the hocks; twist broad and full and running well down with a good covering of soft, silky hair, . . 10

12. TAIL.—Fine, strongly connected with the spine, on a straight line, ... 2

13. UNDERLINE.—As nearly straight as possible; flank low, full and on a straight line with belly and brisket, . . . 4

14. LEGS.—Front legs should be fine boned and stand well apart; knees round and muscular. Hind legs straight, standing well apart, with a muscular hock and a fine boned, flat leg below, ... 4

15. FLESH.—Firm, evenly laid on, and free from lumps or patches, ... 4
16. SKIN.—Moderately thick, not so loose as to separate from the tissue beneath, but at the same time not tight. It should move easily by action of the hand, showing plenty of cellular tissue beneath. The touch or handling qualities should be elastic, mellow (not flabby) and springy. The hair should be close, long, soft, and mossy; the more of it the better, if of the right quality, ... 10

PERFECTION, - - - - 100

POLLED DURHAM CATTLE.

For this breed a Standard of Excellence has not yet been adopted, but for the present, that for Short-horns may be used for judging them, in conjunction with the following registry requirements, adopted by the Polled Durham Breeders' Association, J. H. Miller, Secretary, Mexico, Illinois.

RULE 1.—Animals for registry must be at least one year old.

RULE 2.—Cattle to be eligible to registry must be calved hornless.

RULE 3.—Color and markings characteristic of the Short-horn.

RULE 4.—Animals to be eligible to entry must have seventy-five per cent. or more of Short-horn blood, but where there is a remainder of blood not Short-horn, it must be in part or all of the old native Muley strain.

RULE 5.—Produce of animals already recorded, provided they conform to rules 1, 2 and 3.

RULE 6.—The produce of any bull in the Polled Durham Herd Book, when out of a cow registered in the Short-horn Herd Book, provided they conform to Rules 1, 2 and 3.

RULE 7.—The produce of any cow in Polled Durham registry when by a bull recorded in the American Short-horn Herd Book, provided they conform to Rules 1, 2 and 3.

RULE 8.—After September 4, 1893, no animal shall be registered whose ancestors are not at that time registered, except under Rules 6 and 7, with less than fifteen-sixteenths Short-horn blood, in addition to requirements of Rules 1, 2, 3, and 4.

SUSSEX CATTLE.

Standard of Excellence for Sussex Cattle, compiled from particulars furnished by Overton Lea, of Nashville, Tennessee, owner of the principal herd of Sussex Cattle in the United States, and Secretary of the American Sussex Association, and approved by him.

SCALE OF POINTS FOR SUSSEX CATTLE.

POINTS.	COUNTS.
1. Color,	5
2. Head and Horns,	10
3. Neck,	5
4. Shoulders,	6
5. Chest,	10
6. Heart-girth,	10
7. Ribs,	8
8. Back, Loin and Rump,	15
9. Tail,	2
10. Legs,	4
11. Thighs,	6
12. Skin and Handling,	8
13. Size,	6
14. General Appearance,	5
PERFECTION,	100

DETAILED DESCRIPTION.

POINTS.   COUNTS.

1. COLOR.—Solid red, varying from the light to the very dark red; sometimes the shades are mingled in the same animal, producing a dappled appearance. A little white about the udder is not objectionable in the cow. In both sexes, a few gray or white hairs scattered throughout the coat, sometimes so thick as to be easily noticed in spots, particularly upon the foretop, are regarded with favor rather than otherwise, ... 5

2. HEAD.—Decidedly neat in appearance; nose tolerably wide; muzzle bright, almost golden; thin between the nostrils and the eyes; eye rather prominent; forehead wide;

horns long, longer in cows than in bulls; heavier than in the Devon, and without the Devon's peculiar waxy color; clean and transparent with age, projecting at various angles, similar to those of the Hereford—both with the same angle or curve, however, unless distorted by accident, 10
3. NECK.—Short, strongly set on and clean, ... 5
4. SHOULDERS.—Straight and without any projection at the points, ... 6
5. CHEST.—Wide, open and projecting forward, ... 10
6. HEART-GIRTH.—Full and deep, ... 10
7. RIBS.—Broad and deep, and well sprung, ... 8
8. BACK.—Straight and level from the withers to the setting on of the tail; loin and entire back full of flesh; hips moderately large, but well covered, with a wide space between, and on a level with the back; rump long and slightly sloping. 15
9. TAIL.—Well set on, and dropping perpendicularly, . . 2
10. LEGS.—Of medium length, ... 4
11. THIGHS.—Flat outside and full inside, ... 6
12. SKIN.—Having a mellow touch, with soft and silky coat, . 8
13. SIZE.—Medium weight for bull at three years old, 2,000 pounds; same for cow at three years old, 1,500 pounds, . 6
14. GENERAL APPEARANCE. — Stylish and well bred. The walk square, the step quick, and the head up, . . . 5

PERFECTION, - - - - 100

BROWN SWISS CATTLE.

Standard of Excellence for Brown Swiss cattle, as adopted by the Brown Swiss Cattle Breeders' Association, N. S. Fish, secretary, Groton, Connecticut.

POINTS.	SCALE OF POINTS FOR BULL.	COUNTS.
1.	Head,	2
2.	Face,	2
3.	Ears,	1
4.	Nose and Tongue,	2
5.	Eyes,	1
6.	Horns,	5
7.	Neck,	4
8.	Chest,	4
9.	Back and Loin,	6
10.	Barrel,	8
11.	Hips and Rump,	4
12.	Thighs and Quarters,	4
13.	Legs and Hoofs,	4
14.	Tail and Switch,	4
15.	Hide,	3
16.	Color,	6
17.	Hair,	
18.	and 19 omitted for bulls and heifers,	
20.	Teats,	5
21.	Omitted for bulls and heifers,	
22.	Escutcheon,	7
23.	Disposition,	4
	PERFECTION,	76

POINTS.	DETAILED DESCRIPTION.	COUNTS.
1.	HEAD.—Medium size and rather long,	2
2.	FACE.—Dished; broad between the eyes and narrow between the horns,	2
3.	EARS.—Of a deep orange color within,	1
4.	NOSE.—Black, square, and with the mouth surrounded by a light, meal-colored band, tongue black,	2
5.	EYES.—Full and placid,	1

6. HORNS.—Rather short, flattish and regularly set with black tips, ... 5
7. NECK.—Straight, rather long, and not too heavy at shoulders, ... 4
8. CHEST.—Broad and deep, ... 4
9. BACK.—Level to the setting on of tail and broad across the loin, ... 6
10. BARREL.—Hooped, broad and deep at the flank, . . . 8
11. HIPS.—Wide apart, rump long and broad, ... 4
12. THIGHS.—Wide with heavy quarters, ... 4
13. LEGS.—Short and straight with good hoofs, . . . 4
14. TAIL.—Slender, pliable, not too long, with good switch, . 4
15. HIDE.—Thin and movable, ... 3
16. COLOR.—Shades from dark brown to light brown, and at some seasons of the year gray; slight splashes of white near bag, not objectionable, light stripe along the back, ... 6
17. HAIR.—Between horns light, not reddish, hair on inside of ears light (no points), ...
18.—18 and 19 omitted for bulls and heifers, ...
20. TEATS.—Rather large, set well apart and hanging straight down, ... 5
21.—Omitted for bulls and heifers, ...
22. ESCUTCHEON.—High and broad and full in thighs, . . 7
23. DISPOSITION.—Quiet and good natured, ... 4
               PERFECTION,   - - -   76

POINTS.	SCALE OF POINTS FOR COW.	COUNTS.
1. Head,		2
2. Face,		2
3. Ears,		1
4. Nose and Tongue,		2
5. Eyes,		1
6. Horns,		5
7. Neck,		4
8. Chest,		4
9. Back and Loin,		6
10. Barrel,		8
11. Hips and Rump,		4

BROWN SWISS CATTLE.

12. Thighs and Quarters,	4
13. Legs and Hoofs,	4
14. Tail and Switch.	4
15. Hide,	3
16. Color,	6
17. Hair,	
18. Fore Udder,	10
19. Hind Udder,	10
20. Teats.	5
21. Milk Veins,	4
22. Escutcheon,	7
23. Disposition,	4
PERFECTION,	100

DETAILED DESCRIPTION.

POINTS.   COUNTS.

1. HEAD.—Medium size and rather long, ... 2
2. FACE.—Dished; broad between the eyes and narrow between the horns, ... 2
3. EARS.—Of a deep orange color within, ... 1
4. NOSE.—Black, square, and with the mouth surrounded by a light, meal-colored band; tongue, black, . . . 2
5. EYES.—Full and placid, ... 1
6. HORNS.—Rather short, flattish and regularly set with black tips, ... 5
7. NECK.—Straight, rather long and not too heavy at shoulders, ... 4
8. CHEST.—Broad and deep, ... 4
9. BACK.—Level to the setting on of tail and broad across the loins, ... 6
10. BARREL.—Hooped, broad and deep at flank, . . . 8
11. HIPS.—Wide apart; rump long and broad, . . . 4
12. THIGHS.—Wide, with heavy quarters, ... 4
13. LEGS.—Short and straight, with good hoofs, . . . 4
14. TAIL.—Slender, pliable, not too long, with good switch, . 4
15. HIDE.—Thin and movable, ... 3
16. COLOR.—Shades from dark brown to light brown, and at some seasons of the year, gray; slight splashes of white near udder not objectionable; light stripe along the back, 6
17. HAIR —Between horns light, not reddish; hair on inside of ears light (no points). . . . .
18. FORE UDDER.—Full in form and carried up, reaching far forward on the abdomen, ... 10

19. HIND UDDER.—Not too deeply hung, full in form and well up behind, ... 10
20. TEATS.—Rather large, set well apart and hanging straight down, ... 5
21. MILK VEINS.—Prominent, ... 4
22. ESCUTCHEON.—High and broad and full in thighs, . 7
23. DISPOSITION.—Quiet and good natured, ... 4

     PERFECTION, - - - - 100

In judging bulls and heifers, omit Nos. 18, 19 and 21.

WEST HIGHLAND CATTLE.

Although quite a number of these picturesque and hardy little beef cattle have been sold to come to the States, there is no one at this writing who is known to be maintaining a breeding herd, but as the day is, perhaps, not far distant when there may be several herds in this country, the following detailed description is published:

The head should be beautifully proportioned to the rest of the animal; a fine head, with large tuft of hair on it; the nostrils full; the eyes large and liquid.

The horns should be lengthy, and should come level out of the head, inclining forward and upward; in the cow they should rise up with a graceful slope. Perfection in a cow's horns is of two kinds, according to taste, but some prefer them to come out level from the head, with a peculiar back-set curve and a wider sweep. In the bull the horns should be decidedly strong. The cow's horns rise sooner from the head and are a little longer, preserving their substance and rich color to the very tips.

The neck should be proportionate in length, clean below, and in cows forming a straight line from the head to the shoulders. In point of thickness it should be fully developed, and the bulls should have a crest.

The shoulders should be thick and immensely filled out downward from the point to the lower extremity of the fore-arm.

The back, from the very back of the shoulder, should have a fully rounded development; a hollow behind the shoulder is exceedingly objectionable. Across the hips there should be great breadth; while from the hips backward the quarters should have a very large development, being square betwixt the hips and the tail and betwixt the tail and the hind feet. As in the foreshoulders, the hind thighs should have an immense development.

The tail should be thick and strong, with a full bunch of hair hanging down toward the ground.

The bone, both in the fore and hind legs, should be thick, broad and straight; the hoofs large and well set on, and the legs feathered with hair. There should be great breadth betwixt the forelegs, and the animals should walk with great dignity of motion; indeed, unless an animal possesses this dignified style of carriage, he will have small chances of winning prizes in the show-ring.

The hair should be long, with a graceful wave in it, but a curl in it is a decided fault. The lack of wave in the hair is considered to be a great objection in many of the modern herds, though it is to be accounted for by the growing desire to make Highlanders grow big, and consequently from too careful treatment.

The whole points of the animal have to be considered in the light that he has to make a living in a bare and storm-exposed locality; that, indeed, he has to thrive where a Polled Angus or an Ayrshire would starve.

The question of thickness of skin, where fat, is one which is not left out of consideration; as in other animals, the sweetest beef being, as a rule, that under the thinnest skin. But a West Highlander with too thin a skin would not thrive well on a side of a wind-swept hill.

As a rule the color is black, but fashion now runs on yellow or light duns and on brindles. A well arranged herd should have a mixture of colors, avoiding all those which indicate unhealthy thrivers. A well marked brindle bull is, however, all things being equal, a difficult one to beat at any North British show.

It must always be borne in mind that the Highlanders are one of the most ancient breeds of cattle; that they are a combination of great hardiness with splendid quality of meat, which latter commands the highest price in the principal English markets. At all times they look by far the most noble of the bovine race, whilst their picturesque appearance makes them a handsome addition to the woodland scenery of large estates."

SHEEP
—AND—
GOATS.

How to Judge Wool On the Sheep's Back.

In Henry Stewart's "Shepherd's Manual," page 154, will be found the following excellent advice as to how to judge wool on the sheep's back:

"To determine the evenness or uniformity of the fleece, the shoulder is first examined. Here the finest and best wool should be found. Taking this as the standard, the wool from the ribs, thigh, rump and breech is compared with it; the nearer the latter approaches this in quality, the better. If it is all equal in fineness, the fleece will be "EVEN" in regard to fineness. If the wool on all the parts mentioned is reasonably regular in length, and near to the standard in this respect, the fleece is "EVEN" as regards length of staple. The density is then tested. The hand is closed upon a portion of the rump and on the loin, and if the fleece is found to be as dense and elastic, or springy on those parts as it is at the shoulder, the fleece is "EVEN" as regards density. A perfect fleece will be found of nearly equal fineness from the shoulder to the thigh; of nearly equal length at the shoulder, rib, thigh and back; of equal density on the shoulder and across the loins and free from any of the defects before mentioned."

In August, 1898, in a communication to the "Country Gentleman," Henry Stewart defined the positions of the different qualities of wool upon the sheep, with the following illustration and explanatory notes:

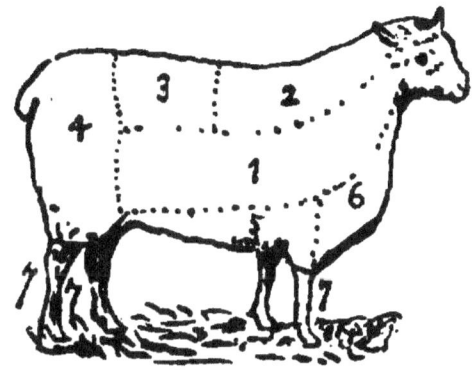

"The wool from the centre of the sides, marked 1, is of the finest quality At 2 and 3 the fleece becomes coarser and shorter, and as the breech is approached, at 4, this coarseness, and length, too, are increased. The most inferior part of the fleece is on the belly and brisket, as at 5 and 6, that on the legs, 7, being almost like hair."

CHEVIOT SHEEP.

Standard of Excellence for Cheviot Sheep, as adopted by the American Cheviot Sheep Breeders Association, R. L. Davidson, Secretary, Cooperstown, N. Y.; also by the National Cheviot Sheep Society, J. A. Guilliams, Secretary, Fincastle, Indiana.

SCALE OF POINTS FOR CHEVIOT SHEEP.

POINTS.	COUNTS.
1. Blood.	15
2. Constitution and Quality,	20
3. Size,	10
4. General Appearance,	10
5. Body,	10
6. Head,	10
7. Neck,	5
8. Legs and Feet,	5
9. Covering,	10
10. Quality of Wool,	5
PERFECTION,	100

DETAILED DESCRIPTION.

POINTS. COUNTS.

1. BLOOD.—Pure bred from one or more importations from Scotland, ... 15

2. CONSTITUTION AND QUALITY.—Indicated by the form of body; deep and large in the breast, and through the heart; back wide and straight, and well covered with lean meat; wide and full in the thigh; deep in the flank; skin soft and pink in color; prominent eyes, healthful countenance; deficiency of brisket or fish back objectionable, . . 20

3. SIZE.—In fair condition, when fully matured, rams should weigh not less than 200 pounds; ewes, 150 pounds (when bred in America). Imported stockrams, 125 to 150 pounds; ewes, 100 to 125 pounds, ... 10

4. GENERAL APPEARANCE.—Good carriage; head well up; elastic movement; showing symmetry of form and uniformity of character throughout, ... 10

CHEVIOT SHEEP.

5. BODY.—Well proportioned; small bone; great scale and length; well finished hindquarters; thick back and loins; standing with legs well placed outside; breast wide and prominent in front; tail wide and well covered with wool, 10
6. HEAD.—Long and broad, and wide between the eyes; ears of medium length and erect; face white, but small black spots on head and ears not objectionable; straight or Roman nose, a white nose objectionable, end of nose dark (but never smut nose on top with black or brown); no tuft of wool on head, ... 10
7. NECK.—Of medium length, thick and well placed on the shoulders, ... 5
8. LEGS AND FEET.—Short legs, well set apart; color, white; no wool on legs; fore legs round, hind legs flat and straight; hoofs black and well shaped, ... 5
9. COVERING.—Body and belly well covered with fleece of medium length and good quality, ... 10
10. QUALITY OF WOOL.—Medium, such as is known in market as half combing wool, ... 5

PERFECTION, - - - 100

COTSWOLD SHEEP.

Standard of Excellence for Cotswold Sheep, as adopted by the American Cotswold Association, George Harding, Secretary, Waukesha, Wisconsin.

SCALE OF POINTS FOR RAM.

POINTS.	COUNTS.
1. Head,	8
2. Face,	4
3. Nostrils,	1
4. Eyes,	2
5. Ears,	4
6. Collar and Neck,	6
7. Shoulders,	8
8. Fore-legs,	4
9. Breast and Girth,	10
10. Fore-flank,	5
11. Back, Ribs and Loin,	12
12. Belly,	3
13. Quarters,	8
14. Hock,	2
15. Twist,	5
16. Fleece,	18
PERFECTION,	100

DETAILED DESCRIPTION.

POINTS.	COUNTS.
1. HEAD.—Not too fine, moderately small, and broad between the eyes and nostrils, but without a short, thick appearance, and in young animals well covered on crown with long lustrous wool,	8
2. FACE.—Either white or slightly mixed with gray, or white dappled with brown,	4
3. NOSTRILS.—Wide and expanded. Nose dark,	1
4. EYES.—Prominent but mild looking,	2
5. EARS.—Broad, long, moderately thin and covered with short hair,	4

6. COLLAR.—Full from breast and shoulders, tapering gradually all the way to where the neck and head join. The neck should be short, thick and strong, indicating con-

stitutional vigor, and free from coarse and loose skin, . 6
7. SHOULDERS.—Broad and full, and at the same join so gradually to the collar forward and chine backward as not to leave the least hollow in either place, ... 8
8. FORE-LEGS.—The mutton on the arm or fore-thigh should come quite to the knee. Leg upright with heavy bone—being clear from superfluous skin, with wool to fetlock, and may be mixed with gray, ... 4
9. BREAST.—Broad and well forward, keeping the legs wide apart. Girth or chest full and deep, ... 10
10. FORE-FLANK.—Quite full, not showing hollow behind the shoulder, ... 5
11. BACK AND LOIN.—Broad, flat and straight, from which the ribs must spring with a fine circular arch, ... 12
12. BELLY.—Straight on underline, ... 3
13. QUARTERS.—Long and full, with mutton quite down to the hock, ... 8
14. HOCK.—Should stand neither in nor out, ... 2
15. TWIST.—Twist or junction inside the thighs, deep wide and full, which, with a broad breast, will keep the legs open and upright, ... 5
16.—FLEECE.—The whole body should be covered with long, lustrous wool, ... 18

             PERFECTION,   -   -   -   - 100

POINTS.	SCALE OF POINTS FOR EWE.	COUNTS.
1. Head,		8
2. Face,		4
3. Nostrils,		1
4. Eyes,		2
5. Ears,		4
6. Collar and Neck,		5
7. Shoulders,		8
8. Fore-legs,		4
9. Breast and Girth,		10
10. Fore-flank,		4
11. Back, Ribs and Loin,		12
12 Belly,		5
13. Quarters,		8
14. Hock,		2

15. Twist, ... 5
16. Fleece, ... 18
                               —
        PERFECTION,   - - - 100

DETAILED DESCRIPTION.

POINTS.   COUNTS.

1. HEAD.—Moderately fine, broad between the eyes and nostrils, but without a short, thick appearance, and well covered on crown with long, lustrous wool, ... 8
2. FACE.—Either white or slightly mixed with gray, or white dappled with brown, ... 4
3. NOSTRILS.—Wide and expanded; nose, dark, . . . 1
4. EYES.—Prominent, but mild looking, ... 2
5. EARS.—Broad, long, moderately thin, and covered with short hair, ... 4
6. COLLAR.—Full from breast and shoulders, tapering gradually all the way to where the neck and head join; the neck should be fine and graceful, and free from coarse and loose skin, ... 5
7. SHOULDERS.—Broad and full, and at the same time join so gradually to the collar forward and chine backward, as not to leave the least hollow in either place, . . . 8
8. FORE-LEGS.—The mutton on the arm or fore-thigh should come quite to the knee; leg upright, with heavy bone, being clear from superfluous skin, with wool to fetlock, and may be mixed with gray, ... 4
9. BREAST.—Broad and well forward, keeping the legs wide apart; girth or chest full and deep, ... 10
10. FORE-FLANK.—Quite full, not showing hollow behind the shoulder, ... 4
11. BACK AND LOIN.—Broad, flat and straight, from which the ribs must spring with a fine circular arch, ... 12
12. BELLY.—Straight on underline, ... 5
13. QUARTERS.—Long and full with mutton quite down to the hock, ... 8
14. HOCK.—Should stand neither in nor out, ... 2
15. TWIST.—Twist or junction inside the thighs, deep, wide and full, which, with a broad breast, will keep the legs open and upright, ... 5
16. FLEECE.—The whole body should be covered with long, lustrous wool, ... 18
                                           —
        PERFECTION,   - - - 100

DORSET-HORN SHEEP.

Standard of Excellence for Dorset-Horn Sheep, as adopted by the Dorset-Horn Sheep Breeders' Association, of America, M. A. Cooper, Secretary, Washington, Pennsylvania.

SCALE OF POINTS FOR DORSET-HORN SHEEP.

POINTS.	COUNTS
1. General Appearance,	20
2. Chest and Brisket,	10
3. Back and Ribs,	15
4. Quarters and Legs,	10
5. Color of Legs and Hoofs,	5
6. Head and Face,	5
7. Neck,	5
8. Horn,	10
9. Foretop and Belly Covering,	10
10. Wool,	10
PERFECTION,	100

DETAILED DESCRIPTION.

POINTS.	COUNTS.
1. GENERAL APPEARANCE.—Head well up, eyes bright and alert, and standing square on legs,	20
2. CHEST AND BRISKET.—Broad full chest, brisket well forward,	10
3. BACK AND RIBS.—Broad straight back, with well sprung ribs,	15
4. QUARTERS AND LEGS.—Heavy square quarters set on short, straight legs, well apart,	10
5. COLOR OF LEGS.—Legs white, with small light colored hoof,	5
6. HEAD AND FACE.—Head small, face white, nostrils well expanded, nose and lips pink in color,	5
7. NECK.—Neck short and round, set well on shoulders,	5
8. HORN.—Horn neat, curving forward, and light in color,	10
9. FORETOP AND BELLY COVERING.—Good foretop and well covered on belly and legs,	10
10. WOOL.—Wool of medium quality and good weight, presenting an even, smooth, white surface,	10
PERFECTION,	100

HAMPSHIREDOWN SHEEP.

Standard of Excellence for Hampshiredown Sheep, as adopted by the Hampshiredown Breeders' Association of America, John I. Gordon, Secretary, Mercer, Pa. [Slightly changed in arrangement for this publication].

SCALE OF POINTS FOR HAMPSHIREDOWN SHEEP.

POINTS.	COUNTS.
1. Head,	5
2. Ears and Eyes,	3
3. Color of Head and Legs,	4
4. Legs,	3
5. Neck,	5
6. Shoulders,	10
7. Chest,	15
8. Back and Loin,	20
9. Quarters,	25
10. Wool,	10
PERFECTION,	100

DETAILED DESCRIPTION.

POINTS. COUNTS.

1. HEAD.—Moderately large but not coarse; well covered with wool on forehead and cheeks; nostrils wide, . . . 5
2. EARS AND EYES.—Ears moderately long and thin, and dark brown or black color; eyes prominent and lustrous, . . . 3
3. COLOR OF HEAD AND LEGS.—Dark brown or black, . . 4
4. LEGS.—Well under outside of body; straight, with good size of bone; black, ... 3
5. NECK.—A regular taper from shoulders to head, without any hollow in front of shoulders; set high up on body, . . 5
6. SHOULDERS.—Sloping; full, and not higher than the line of back and neck, ... 10
7. CHEST.—Deep and full in the heart place, with breast prominent and full, ... 15
8. BACK AND LOIN.—Back, straight with full spring of rib; loin, wide and straight, without depression in front of hips, ... 20
9. QUARTERS.—Long from hip to rump, without sloping, and

deep in the thigh, ... 10. Broad in hips and rump, with full hams, . . 10. Inside of thighs full, . . 5, . . 25
10. WOOL.—Forehead and cheeks, . . . 2. Belly well covered, . . 3. Quality, . . 5, ... 10

      PERFECTION, - - - 100

HIGHLAND BLACK-FACED SHEEP.

Whilst quite a number of these hardy little sheep have been brought to the United States, there has not so far been formed any Association of Breeders of them similar to those who watch the interests of other breeds. The following description from Henry Stewart's "Shepherd's Manual" will therefore serve, at present, in place of a "Standard of Excellence":

They are a horned breed, the horns of the ram being massive, and spirally curved. The face is black, with a thick muzzle; the eye is bright and wild; the body square and compact, with good quarters and a broad saddle. They are very muscular and active, and remarkably hardy, able to endure the privations incident to a life of continual exposure upon bleak and storm-beaten mountains. Only the heaviest snow-drifts, followed by thawing, freezing and crusting of snow, overcome them. The mutton of this breed is of peculiarly fine flavor, and the saddles are in great request. The carcass weighs about 65 pounds, and the fleece averages about 8 pounds of washed wool. The breed improves easily under the care of a judicious breeder, but the natural qualities of this sheep are such that it is fitted for a place where no others would profitably thrive, and a change in its character that would cause it to lose this quality would unfit it for its position, and deprive it of its chief value. How vast the room in our exposed mountain localities, or on our unsheltered northern plains for such a sheep as this; a race hardy and self-dependent, and that would produce choice mutton, and a fleece well adapted for rural manufactures of coarse cloths, carpets, blankets and rugs."

LEICESTER SHEEP.

A Standard of Excellence for Leicester Sheep has not yet been compiled in the United States, but the following description of the breed is taken from Vol. I. of the American Leicester Record, published in 1898, by The American Leicester Breeders' Association, A. J. Temple, Secretary. Cameron, Illinois.

"The Leicester has been bred in Scotland and the border counties of England for more than a century and consequently is not a "fleeting thing of a day," but a distinct breed, and for purity of breeding can compare favorably with any other breed of sheep. They are becoming more popular each year in the United States and Canada. One point of their popularity being the ready sale of rams at good prices for use in crossing on other breeds to produce early lambs for the market. The Leicesters mature very early and are of a good size; the rams weighing 250 to 300 pounds, and the ewes 200 to 250 pounds each, fine bone and very little offal, making them a profitable animal for market as well as for wool. The average weight of fleece is 10 to 15 pounds. Wool, 10 to 12 inches long, glossy, of firm fibre, and is conceded to be the best species of long or combing wool.

In appearance the Leicester is a fine looking animal, white in the face, eyes clear and prominent, ears well set and free from blue. Sometimes black spots appear on the ears, but are considered no disadvantage. The neck is set well into the shoulder, full and broad at the base; the shoulders deep and wide, breast full and broad and no uneven or angular formation where the shoulder joins the neck or back; deep in flank, quarters long and square. The legs are bare, being covered with hair rather than wool, and stand wide apart with no looseness of skin on them, bone fine and hard, legs of moderate length; straight on back and broad, light in the belly, nearly as straight below as above (showing light offal), noble bearing, style and action, and the best appearing of any of the long wooled breeds."

LINCOLN SHEEP.

Standard of Excellence for Lincoln Sheep, as adopted by The National Lincoln Sheep Breeders' Association, H. A. Daniels, Secretary, Elva, Michigan.

SCALE OF POINTS FOR LINCOLN SHEEP.

POINTS.	COUNTS.
1. Constitution,	25
2. Size,	10
3. Appearance,	10
4. Body,	15
5. Head,	10
6. Neck,	5
7. Legs,	10
8. Fleece,	10
9. Quality of Wool,	5
PERFECTION,	100

DETAILED DESCRIPTION.

POINTS. COUNTS.

1. CONSTITUTION.—Body deep, back wide and straight; wide and full in the thigh; bright, large eyes; skin soft and of a pink color, ... 25

2. SIZE.—Matured rams not less than 250 pounds when in good condition. Matured ewes not less than 200 pounds, . . 10

3. APPEARANCE.—Good carriage and symmetry of form, . . 10

4. BODY.—Well proportioned, good bone and length; broad hindquarters; legs standing well apart; breast wide and deep, ... 15

5. HEAD.—Should be covered with wool to the ears; eyes expressive; ears fair length; dotted or mottled in color, . 10

6. NECK.—Medium length; good muscle; well set on body, . 5

7. LEGS.—Broad and set well apart; good shape; color white, but some brown spots do not disqualify; wooled to the knees, ... 10

8. FLEECE.—Of even length and quality over body; not less than eight inches long for one year's growth, . . . 10
9. QUALITY OF WOOL.—Rather fine, long wool; strong, lustrous fiber; no tendency to cot, ... 5

PERFECTION, - - - - 100

[The American Lincoln Breeders' Association, of which Lyman C. Graham, of Cameron, Illinois, is secretary, has not yet adopted a "Standard of Excellence."—ED.]

American Rambouillet Merino Sheep.

Mr. L. G. Townsend, of Ionia, Michigan, Secretary of the American Rambouillet Sheep Breeders' Association, writes the editor that a committee of breeders of this valuable breed is preparing a scale of points and standard of excellence for adoption by the association, and that it will probably be ready in a few months.— In the meantime, the following description, published in the American Rambouillet Record, from the pen of Mr. Thomas Wyckoff, a prominent breeder, and one of the Board of Directors of the association, will be a safe guide:

CHARACTERISTICS OF RAMBOUILLET SHEEP AND THEIR WOOL.

AMERICAN RAMBOUILLET SHEEP should have large frame, large, strong bone, well rounded and symmetrical bodies, well up on legs, bright pink skin, always plain and free from wrinkles. Broad head, bright eyes, quick movement, broad backs and broad chests are indispensable.

These sheep have long been noted as a mutton sheep, their fine juicy flesh having no superior.

They are noted for their early maturity and quick-feeding properties, being fully equal to the down breeds in this respect.

The rams are usually well horned, but not always, and weigh at maturity from 175 to 250 pounds.

The ewes are noted as good mothers, heavy milkers, one-half usually producing twins. They weigh 110 to 150 pounds. Wethers attain 150 to 200 pounds.

Being strong, vigorous and healthy, their impressive power is very great, and they are not liable to constitutional break-down in service.

They will bear herding in large flocks, and their great hardiness permits them to stand all kinds of weather without housing.

Their fullness of carcass, brightness of look, length of body, vigor of carriage and great strength, make them excellent and reliable re-producers, and quick, healthy feeders on the range.

Rambouillet wool is of the finest quality, has a beautiful crimp, is usually white, sometimes of a buff color, very compact, opens in large layers, has just yolk enough to promote a rapid and vigorous growth, shows no crust formation, seldom any jar hairs, and is always noted for its length, strength and elasticity.

It is from three to five inches in length, often six and one-half inches for one year's growth.

Rambouillet sheep should be well wooled to the feet and to the nose.

Rams shear annually from 12 to 24 pounds; ewes from 6 to 10 pounds.

The wool scours from 50 to 55 per cent. for the manufacturer, and no other Merino wool shows so deep a staple.

"No finer wool can be produced."

The fineness, length, soundness of staple, and remarkable freedom from grease have brought these sheep into deserved favor.

American wools having this standard of excellence cannot fail to be in demand.

Black-Top Spanish Merino Sheep.

Standard of Excellence for Black-Top Spanish Merino Sheep, as adopted by The Black-Top Spanish Merino Sheep Breeders' Publishing Association. W. G. Berry, Secretary, Houstonville, Washington county, Pennsylvania.

SCALE OF POINTS FOR BLACK-TOP SPANISH MERINO SHEEP.

POINTS. COUNTS.
1. Blood,
2. Constitution, ... 15
3. Size, ... 12
4. General Appearance, ... 8
5. Body, ... 15
6. Head, ... 5
7. Neck, ... 4
8. Legs and feet, ... 10
9. Covering, ... 8
10. Quality, ... 7
11. Density, ... 7
12. Length, ... 8
13. Oil, ... 6

PERFECTION, - - - 100

DETAILED DESCRIPTION.

POINTS. COUNTS.
1. BLOOD.—Purely bred from the Humphrey importation of Merino sheep from Spain, in the year 1802, as bred by W. R. Dickinson, of Steubenville, Ohio, ...
2. CONSTITUTION.—Indicated by physical development; deep and large in the breast and through the heart; broad back; very heavy, square quarters; skin of fine texture, and pinkish in color; expansive nostril; brilliant eye; healthful countenance; and good feeders, . . . 15
3. SIZE.—In good condition, with fleece of five months' growth, full grown rams should weigh not less than 175 pounds, and ewes not less than 130 pounds, ... 12
4. GENERAL APPEARANCE.—Head carried well up; standing squarely on feet and legs; well rounded body, showing in all points symmetry of form, ... 3

5. BODY.—Throughout, heavy boned; well proportioned in length; smooth joints; ribs starting horizontally from the back-bone, and well rounded to breast-bone; breast-bone wide, strong and prominent in front; strong, straight and heavy back-bone; heavy, muscular quarters, deep through and squarely formed before and behind; shoulders broad and flat, and not projecting sharply above the back-bone; muscles firm and heavy, and body entirely free from folds. There may be a slight throatiness, and a small dew-lap—smaller on the ewes than on the rams, ... 15

6. HEAD.—Wide, medium in length; eyes clear and bright; prominent ears, medium in size and covered with soft fur. Ewes should give no appearance of horns, while upon the rams the horns should be well developed, clear in color, and symmetrically curved, without tendency to extreme expansion, ... 5

7. NECK.—Medium in length and very heavy, especially with the rams, deepening toward the shoulder, ... 4

8. LEGS AND FEET.—Legs medium in length, set well apart, medium bone and smooth joints. The feet must be well shaped, medium sized, firm and solid, ... 10

9. COVERING.—Evenness of fleece and crimp; body and legs covered to the knees; head covered forward between the eyes; the surface should be free from hair or gare, . 8

10. QUALITY.—Medium or fine, such as is known in the market as fine delaine, ... 7

11. DENSITY.—Shown by compactness of fleece, which should open freely, and have no tendency to be stringy or knotty, 7

12. LENGTH.—At twelve months, growth must be not less than three inches, and as near as may be of uniform length, . 8

13. OIL.—Evenly distributed, white, soft and flowing freely from skin to surface, forming on the exterior a uniform dark coating, ... 6

PERFECTION, - - - - 100

Improved Black-Top Merino Sheep.

Standard of Excellence for Improved Black-Top Merino Sheep, as adopted by The Improved Black-Top Merino Association, L. M. Crothers, Secretary, Crothers Washington county, Pennsylvania.

SCALE OF POINTS FOR IMPROVED BLACK-TOP MERINO SHEEP.

POINTS.	COUNTS.
1. Constitution,	16
2. Size,	14
3. General Appearance,	8
4. Body,	16
5. Head,	4
6. Neck,	3
7. Legs,	9
8. Covering,	8
9. Quality of Wool,	7
10. Fleece,	6
11. Staple,	8
12. Oil,	6
PERFECTION,	100

DETAILED DESCRIPTION.

POINTS.	COUNTS.
1. CONSTITUTION.—	16
2. SIZE.—Rams shall weigh at maturity 180 pounds; ewes 130 pounds,	14
3. GENERAL APPEARANCE,—	8
4. BODY.—Large, well proportioned and symmetrical in all its parts,	16
5. HEAD.—Medium in size, well carried up; wool extending forward between the eyes,	4
6. NECK.—Short and well shaped,	3
7. LEGS.—Short, set well apart, with smooth joints and small, thin, shelly feet,	9

8. COVERING.—An even fleece, beautifully crimped, covering the body and legs to the knees, and extending well forward between the eyes, ... 8
9. QUALITY OF WOOL.—Medium or fine delaine, . . 7
10. FLEECE.—Compact and even quality, ... 6
11. STAPLE.—A year's growth should not be less than three and one-half inches, ... 8
12. OIL.—Evenly distributed, flowing to the surface and forming a uniform dark or black top, ... 6

    PERFECTION, - - - - 100

National Delaine Merino Sheep.

Standard of Excellence for National Delaine Merino Sheep, as adopted by The National Delaine Merino Sheep Breeders' Association, John C. McNary, Secretary, Canonsburg, Pennsylvania.

SCALE OF POINTS FOR NATIONAL DELAINE MERINO SHEEP.

POINTS.	COUNTS.
1. Constitution,	10
2. Heavy Around the Heart,	6
3. Short, Heavy Neck,	6
4. Good Dewlap,	5
5. Broad Back,	8
6. Well-sprung Ribs,	5
7. Short Legs,	6
8. Heavy Bone,	8
9. Small, Sharp Foot,	10
10. Length of Staple, 1 Year's Growth, 3 Inches,	8
11. Density of Fleece,	8
12. Darkish Cast on Top,	5
13. Opening up White,	5
14. Good Flow of White Oil,	5
15. Good Crimp in Staple,	5
Perfection,	100

Weight of rams at maturity not less than 150 pounds. Weight of ewes at maturity not less than 100 pounds.

DETAILED DESCRIPTION.

POINTS. COUNTS.

1. CONSTITUTION, ... 10

Requisites—Robust; eyes bright; compactly built; head and neck on line with the back.

Objections—A dull, languid look; loose, slabby make; slim neck; low on top; long, narrow face and curving back.

2. HEAVY AROUND THE HEART, ... 6

Requisites—The entire chest uniform in size, deep and rounding.

Objections—Pot-gutted; flat-shouldered; narrow between the forelegs, and high hip bones.

3. SHORT, HEAVY NECK, ... 6
    Requisites—Straight and short from the top of the head to shoulder; deep and folded.
    Objections—Long, round neck; concave on top and smooth.

4. GOOD DEWLAP, ... 5
    Requisites—Starting on top of neck near the shoulder and widening on each side and hanging deep in front of the forelegs, with small folds at intervals to back of head.
    Objections—Neck tucked up and narrow in front of forelegs; and back of the jaws round and straight on under side.

5. BROAD BACK, ... 8
    Requisites—Straight and flat from shoulder to tail.
    Objections—Sharp on top; curving and drooping from hip bones to tail.

6. WELL-SPRUNG RIBS, ... 5
    Requisites—Starting at right angles from backbone; curving and deep, as long before as behind, making a barrel-like chest.
    Objections—Starting low and downward; flat and shorter before than behind.

7. SHORT LEGS, ... 6
    Requisites—Straight; short and flat boned; standing fair under the body.
    Objections—Long; crooked; slender; drawn together; trying to stand upon the least ground possible.

8. HEAVY BONE, ... 8
    Requisites—Flat, broad-limbed; strong ribs; heavy shoulder blade.
    Objections—Round, small limbs; narrow ribs.

9. SMALL, SHARP FOOT, ... 10
    Requisites—Neat foot, well under the leg; wide at heel.
    Objections—A clubby foot, growing long in toes; narrow and close at heel; large foot joints, and standing forward of the legs.

10. LENGTH OF STAPLE, ... 8

Requisites—Uniform length all over the body, belly and limbs to the knees, and covering the face square to the eyes.

Objections—Growing shorter on sides and belly, bare on legs and face.

11. DENSITY OF FLEECE, ... 8

Requisites—The fleece presenting a smooth, uniform surface.

Objections—Opening up along the back, hanging in strings on shoulder, bare between the legs, and from the knees down, with a thin, open, light fleece.

12. DARKISH CAST ON TOP, ... 5

Requisites—Uniformly dark on outer end of wool.

Objections—Black, crusty top along the back, white and bare along the sides and belly.

13. OPENING UP WHITE. 5

Requisites—Pure, soluble white oil evenly distributed along the fiber.

Objections—Yellow, gummy, curdled oil, causing a black, crusty top, will not dissolve in washing, leaving the fleece when shorn, yellow and unsalable.

14. GOOD FLOW OF WHITE OIL, ... 5

Requisites—Oil enough to protect the fleece, giving it a healthy and rich appearance.

Objections—Oil in excess of wool.

15. GOOD CRIMP IN STAPLE, ... 5

Requisites—Short spiral crimp, evidence of pure Merino wool,

Objections—Coarse, harsh, stringy fiber, evidence of mixed or impure blood.

PERFECTION, - - - - 100

National Dickinson Merino Sheep.

Standard of Excellence for National Dickinson Merino Sheep, as adopted by The National Dickinson Record Company, H. G. McDowell, Secretary, Canton, Ohio. [Slightly changed in arrangement for this publication.]

SCALE OF POINTS FOR NATIONAL DICKINSON MERINO SHEEP.

POINTS.	COUNTS.
1. Blood,	
2. Body,	
3. Skin,	4
4. Head,	4
5. Nose,	8
6. Ears,	3
7. Horns,	3
8. Neck,	4
9. Shoulders,	5
10. Back,	8
11. Loins,	3
12. Hips,	3
13. Thighs,	4
14. Limbs,	5
15. Hoofs,	4
16. Size,	5
17. Internal Organs,	4
18. Maturity,	3
19. Density of Fleece,	6
20. Staple,	4
21. Quality,	6
22. Quantity,	6
23. Covering,	8
24. Oil,	5
Perfection,	100

DETAILED DESCRIPTION.

POINTS.   COUNTS.

1. BLOOD.—Tracing their descent to the standard bred flock of James McDowell, of Canton, Stark county, Ohio, (without admixture of impure blood), which flock descends directly from the thoroughbred flock of W. R. Dickinson, of Ohio, which were purely bred from Merino sheep imported from Spain to the United States by David Humphreys, of Derby, New Haven county, Connecticut, in the year 1802.

2. BODY.—Deep, round, wide and long, showing mutton capacity, good feeding and thriving qualities; heavy, thick

flesh; straight under the top lines, well proportioned, filling every part of its skin when fully matured.

3. SKIN.—Thick, soft, not raised in corrugations, pink red, . 4
4. HEAD.—Small, carried high; quiet, placid eye, . . . 4
5. NOSE.—White, not mottled, covered with fine, soft white, hair, wide and slightly arched, ... 3
6. EARS.—Short, thick; covered with fine glossy hair, . . 3
7. HORNS.—Small, neatly curved, light yellow color; *better without any horns*, ... 3
8. NECK.—Short, arched under and on top; the base very strong, ... 4
9. SHOULDERS.—Wide, deep, rounded; breast bone projecting forward of front limbs, ... 5
10. BACK.—Straight, wide, ribs extending out horizontal from spinal column, rounding in line with shoulders extending close back to hips, ... 8
11. LOINS.—Strong, wide, ... 3
12. HIPS.—Wide, long, ... 3
13. THIGHS.—Wide, thick; flesh extending close down to hock joints, ... 4
14. LIMBS.—Short, bone heavy; joints smooth and flat, the contour of to show perpendicular lines from elbow and stifle joints to center of hoofs, and from base of tail to center of a straight line drawn horizontally from caps of hock joints, when standing erect on limbs, . . . 5
15. HOOFS.—Deep, thin white texture, tough and elastic, . . 4
16. SIZE.—Full-grown rams 200 pounds, and ewes, 150 pounds, . 5
17. INTERNAL ORGANS.—Strong, ... 4
18. MATURITY.—Mature early, two and one-half years, . . 3
19. DENSITY OF FLEECE.—Smooth, even, dense soft to the touch, ... 6
20. STAPLE.—Three to five inches, fibers glossy, crimped, . . 4
21 —QUALITY.—XX, XXX or above, fine Delaine combing, . 6
22. QUANTITY.—Rams, 15 to 25 pounds; ewes, 10 to 15 pounds unwashed wool, ... 6
23. COVERING.—Entire body covered with even length and grade of wool, except parts injuring thrift and comfort of the sheep, entirely free from gum and hair. . . . 8
24. OIL.—Very fluid, white or nearly so, enough to preserve the wool, raising to outer ends of fibers, ... 5

PERFECTION, - - - - 100

National Improved Saxony Merino Sheep.

Standard of Excellence for National Improved Saxony Sheep, as adopted by The National Improved Saxony Sheep Breeder's Association, John G. Clark, secretary, Toledo, Pa.

SCALE OF POINTS FOR NATIONAL IMPROVED SAXONY SHEEP.

POINTS.	COUNTS.
1. Blood,	1
2. Constitution,	15
3. Size,	10
4. Body,	12
5. Head,	5
6. Neck,	5
7. Legs and Feet,	5
8. Evenness of Fleece,	15
9. Density of Fleece,	12
10. Staple,	10
11. Oil,	10
PERFECTION,	100

Only the three highest grades, Picknic, Picklock, and XXX are admitted. Every sheep not making XXX is rejected.

DETAILED DESCRIPTION.

POINTS.	COUNTS.
1. BLOOD.—Tracing through some of the best flocks to imported stock, and the wool must grade XXX or above,	1
2. CONSTITUTION.—Indicated by general appearance,	15
3. SIZE,	10
4. BODY.—Well proportioned and free from wrinkles,	12
5. HEAD,	5
6. NECK.—Short, well set, only slight dewlap,	5
7. LEGS AND FEET.—Legs short and heavy boned,	5
8. EVENNESS OF FLEECE.—Well covered on belly, face and legs,	15

NATIONAL IMPROVED SAXONY MERINO SHEEP.

   9. DENSITY OF FLEECE, ... 12
10. LENGTH OF STAPLE.—And fine crimp, ... 10
11. OIL.—Wool opening white, ... 10

                      PERFECTION, - - - - 100

The Improved Saxony Sheep should be large, strong, heavy boned, well proportioned, compactly built, free from wrinkles or folds, short, well-set neck with only slight dewlap, good carriage, stylish, large girt around the heart, and well-shaped feet. The wool must grade XXX or above, long, white, dense crimpy, free from curly spots on top of shoulders or back, and evenly over the whole body.

Standard American Merino Sheep.

Standard of Excellence for Standard American Merino Sheep, as adopted by the Standard American Merino Sheep Breeders' Association, John P. Ray, Secretary, Hemlock Lake, N. Y., and republished by his permission. [Slightly changed in arrangement for this publication.]

SCALE OF POINTS FOR STANDARD AMERICAN MERINO SHEEP.

POINTS.   COUNTS.
- A. Constitution, ... 15
- B. Form, ... 40
- C. Wrinkles, ... 15
- D. Density of Fleece, ... 15
- E. Covering, ... 15

                PERFECTION,   - - - 100

DETAILED DESCRIPTION.

POINTS.   COUNTS.

CONSTITUTION—Fifteen Points.

1. Bone, ... 5
2. Physical development and general appearance, . . . 10

FORM—Forty Points.

3. A broad head, broad, wrinkly nose and face, covered with a soft velvety coat, ... 5
4. Short, broad, muscular neck, well set on shoulders, . . 5
5. Massiveness of shoulder, as to depth and breadth, . . 5
6. Level, straight back and rotundity of rib, . . . 5
7. Breadth and length of hips, ... 5
8. Straight forelegs, well set apart, ... 8
9. Straight hind legs, and set so as to give a perpendicular appearance to hind parts, ... 5
10. Soft, thick, velvety ear, ... 2
11. Pure white nose, ears and hoofs, ... 5

WRINKLES—Fifteen Points.

12. Heavy, pendulous neck, ... 5
13. Across arm and point of shoulder on side, and running well under, ... 5

14. Tail, hip-folds and flank, ... 5

MODIFIED FOR DELAINE RAM YIELDING A STAPLE OF 2¼ INCH AND UPWARD.

12. A deep gullet and heavy cross at brisket, ... 5
13. Heavy flank with fold extending upward on side and back of shoulder, ... 5
14. Heavy tail, ... 5

DENSITY OF FLEECE—Fifteen Points.

15. On neck, ... 3
16. On back, ... 3
17. On side, ... 3
18. On hip and extending to flank, ... 3
19. On belly, ... 3

COVERING—Fifteen Points.

20. Crown of head or cap, ... 3
21. Cheek, ... 2
22. Fore leg, ... 2
23. Arm pit, ... 2
24. Hind leg, ... 2
25. Inside of flank, ... 3
26. Connection between tag wool and belly, ... 1

PERFECTION, - - - - 100

Fibre to be indicated as "fine," "medium" and "coarse"; oil, as "buff" and "white."

Standard Delaine Spanish Merino Sheep.

Standard of Excellence for Standard Delaine Spanish Merino Sheep, as adopted by the Standard Delaine Spanish Merino Sheep Breeders' Association, S. M. Cleaver, Secretary, East Bethlehem, Washington county, Pa.

SCALE OF POINTS FOR STANDARD DELAINE SPANISH MERINO SHEEP.

POINTS.	COUNTS.
1. Blood,	
2. Constitution,	20
3. Fleece,	10
4. Density of Fleece,	3
5. Evenness of Surface,	3
6. Evenness of Crimp,	3
7. Length of Fiber,	2
8. Oil,	9
9. Head,	4
10. Eyes,	3
11. Nose,	4
12. Ears,	2
13. Neck,	4
14. Covering and Skin,	4
15. Legs,	2
16. Feet,	4
17. Quarters and Back,	10
18. Weight,	8
19. General Appearance,	5
PERFECTION,	100

Any sheep scaling below 60 per cent. in any point cannot be recorded.

DETAILED DESCRIPTION.

POINTS. COUNTS.

1. BLOOD.—Pure Merino blood, which must be established by certificate,
2. CONSTITUTION.—Indicated by a deep chest, long rib well arched, giving heart and lung room, with great digestive capacity, ... 20
3. FLEECE.—Fleece XX and Delaine wool. This includes the quantity and quality as shown by weight of fleece, the

length and strength of staple, crimp, fineness and trueness of fiber, ... 10
4. DENSITY OF FLEECE, ... 3
5. EVENNESS OF SURFACE, ... 3
6. EVENNESS OF CRIMP, ... 3
7. LENGTH OF FIBER, ... 2
8. OIL.—Free flowing oil of the best quality and the right quantity to protect the sheep and preserve the fleece, . . 9
9. HEAD—Head medium size. Ewes showing a feminine appearance; Rams, a masculine, with properly turned horns, 4
10. EYES.—Eyes bright, prominent and well set apart, with a thick, soft eyelid, ... 3
11. NOSE.—Nose short, broad, with well expanded nostrils, skin thick and covered with a thick furry coating, joining the wool 1 inch below the eyes, ... 4
12. EARS—Ears medium size, set well apart, thickly coated, . 2
13. NECK.—Neck short on top, deep and strongly attached to shoulders, tapering to head; Rams with a fold across the breast, and deep neck, ... 4
14. COVERING AND SKIN.—Fleece covering over the entire body, head and legs; skin thick and spungy, ... 4
15. LEGS.—Legs short, strong and well apart, ... 2
16. FEET.—Feet neatly shaped, thin hoof, well set under the leg, ... 4
17. QUARTERS AND BACK.—Quarters, deep and well rounded; back broad, straight and strongly coupled to quarters, . . 10
18. WEIGHT.—Weight of ewes at maturity, 100 pounds and above; Rams, 150 and above, ... 8
19. GENERAL APPEARANCE.—General appearance, good carriage, bold and vigorous style, symmetrical form, . . 5

PERFECTION, - - - - 100

Other Merino Associations.

The following associations have not adopted standards of excellence for Merinos, viz:

The National Merino Sheep Register Association, R. O. Logan, Secretary, California, Michigan.

The New York State American Merino Sheep Breeders' Association, J. Horatio Earll, Secretary, Skaneateles, N. Y.

The Ohio Spanish Merino Sheep Breeders' Association, F. C. Stanley, Secretary, Edison, Ohio.

The Vermont Atwood Club Register, Geo. Hammond, Secretary, Middlebury, Vt.

The Vermont Merino Sheep Breeders' Association, C. A. Chapman, Ferrisburg, Vt.

OXFORD DOWN SHEEP.

Standard of Excellence for Oxford Down Sheep, as adopted by the American Oxford Down Sheep Record Association, W. A. Shafor, Secretary, Middletown, Ohio.

SCALE OF POINTS FOR OXFORD DOWN SHEEP.

POINTS.	COUNTS.
1. Head,	8
2. Face,	4
3. Nostrils.	1
4. Eyes,	1
5. Ears,	4
6. Collar,	6
7. Shoulder,	8
8. Fore legs,	4
9. Breast,	10
10. Fore flank,	5
11. Back and Loin,	12
12. Belly,	8
13. Quarters,	8
14. Hock,	2
15. Twist or Junction,	6
16. Fleece,	18
PERFECTION,	100

DETAILED DESCRIPTION.

POINTS. COUNTS.

1. HEAD.—Not too fine, moderately small and broad between the eyes and nostrils, but without a short, thick appearance; crown well covered with good wool, . . . 8

2. FACE.—Either brown or gray, but not speckled or white; with a white or gray spot on end of nose, . . . 4

3. NOSTRILS.—Wide and expanded, and dark, . . . 1

4. EYES.—Prominent, but mild, . . . 1

5. EARS.—Broad, moderately long, thin, and covered with short brownish hair or wool, . . . 4

6. COLLAR.—Full from breast and shoulders, tapering gradually all the way to where the head and neck join; the

neck short, thick and strong (with masculine appearance in rams), indicating constitutional vigor, and free from coarse or loose skin, ... 6
7. SHOULDER.—Broad and full, and at the same time join so gradually to the collar forward and the chine backward as not to leave the least hollow in either place. . . . 8
8. FORE LEGS.—The mutton on the arm or forethigh should come quite to the knee; leg heavy bone and upright, being clear from superfluous skin; dark brown or smoky in color; should stand square and well apart, ... 4
9. BREAST.—Broad and well forward, keeping legs well apart; girth or chest full and deep, ... 10
10. FORE FLANK.—Quite full, not showing hollow behind shoulder, ... 5
11. BACK AND LOIN.—Broad, flat and straight, from which the ribs must spring with a fine circular arch, ... 12
12. BELLY.—Straight on underline, ... 8
13. QUARTERS.—Long and full, with mutton quite down to the hock, ... 8
14. HOCK.—Stand neither in nor out, but straight, . . . 2
15. TWIST OR JUNCTION.—Inside the thigh deep, wide and full, which with a broad breast, will keep the legs open and upright, ... 6
16. FLEECE.—The whole body should be covered with wool of a close texture, a good length, and fine quality, . . . 18

PERFECTION, - - - - 100

SHROPSHIRE SHEEP.

Standard of Excellence for Shropshire Sheep, as adopted by the American Shropshire Registry Association, Mortimor Levering, Secretary, La Fayette Indiana.

SCALE OF POINTS FOR SHROPSHIRE SHEEP.

POINTS.	COUNTS.
1. Constitution,	25
2. Size,	10
3. General Appearance,	10
4. Body,	15
5. Head,	10
6. Neck,	5
7. Legs and Feet,	10
8. Fleece,	10
9. Quality of Wool,	5
PERFECTION,	100

DETAILED DESCRIPTION.

POINTS.   COUNTS.

1. CONSTITUTION.—And quality indicated by the form of body; deep and large in breast and through the heart, back wide, straight and well covered with lean meat or muscle; wide and full in the thigh, deep in flank; skin thick but soft and of a pink color; prominent, brilliant eyes and healthful countenance, ... 25

    Objections.—Deficiency of brisket, light around the heart, fish back, pointed shoulders, tucked in flank, pale or too dark skin objectionable.

2. SIZE.—In fair condition when fully matured, rams should weigh not less than 225 pounds, and ewes not less than 175 pounds, ... 10

    Objections—Rams in full flesh 175 pounds or under; ewes in full flesh 150 pounds or under.

3. GENERAL APPEARANCE.—And character, good carriage; head well up; elastic movement, showing great symmetry of form and uniformity of character throughout, . . 10

SHROPSHIRE SHEEP.

Objections—Head drooping, low in neck, sluggish movement.

4. BODY.—Well proportioned, medium bones, great scale and length, well finished hind-quarters, thick back and loins, twist deep and full, standing with legs well placed outside, breast wide and extending well forward, . . . 15

Objections—Too fine bones, short body, deficient in twist, legs close together, light in brisket.

5. HEAD.—Short and broad; wide between the ears and between the eyes; short from top of head to tip of nose; ears short of medium size; eyes expressive; head should be well covered with wool to a point even with the eyes, without any appearance of horns; color of face dark brown, ... 10

Objections—Horns disqualify, white face disqualifies, head with prominent bones, bare on top of head.

6. NECK.—Medium length, good bone and muscular development; and especially with the rams heavier toward the shoulders, well set high up, and rising from that point to the back of the head, ... 5

7. LEGS AND FEET.—Broad, short, straight, well set apart, well shaped; color dark brown, and well wooled to the knees, ... 10

8. FLEECE.—Body, head, belly and legs to knees well covered with fleece of even length and quality; scrotum of rams well covered with wool, ... 10

9. QUALITY OF WOOL.—Medium, such as is known in our markets as "medium-delaine" and "half-combing wool" strong, fine, lustrous fiber, without tendency to mat or felt together, and at one years' growth not less than three and one-half inches in length, ... 5

PERFECTION, - - - - 100

SOUTHDOWN SHEEP.

Standard of Excellence for Southdown Sheep, as adopted by The American Southdown Association, John G. Springer, Secretary, Springfield, Illinois.

SCALE OF POINTS FOR SOUTHDOWN SHEEP.

POINTS.	COUNTS.
1. Head,	5
2. Lips,	1
3. Ears,	2
4. Eyes,	3
5. Face,	3
6. Neck,	4
7. Shoulders,	5
8. Breast,	5
9. Back and Loin,	7
10. Ribs,	6
11. Rump,	6
12. Hips,	6
13. Thighs,	6
14. Limbs,	3
15. Forelegs,	2
16. Hindlegs,	2
17. Belly and Flank,	5
18. Fleece,	12
19. Form,	9
20. General Appearance,	8
PERFECTION,	100

DETAILED DESCRIPTION.

POINTS.	COUNTS.
1. HEAD.—Head medium in size and hornless, fine, carried well up; the forehead or face well covered with wool, especially between the ears and on the cheeks, and in the ewe slightly dished,	5
2. LIPS.—Lips and under jaw fine and thin,	1
3. EARS.—Ears rather small, tolerably wide apart, covered with fine hair, and carried with a lively back and forth movement,	2
4. EYES.—Eyes full and bright,	3

5. FACE.—Face a uniform tint of brown, or gray, or mouse color, ... 8
6. NECK.—Neck short, fine at the head, but nicely tapering, and broad and straight on top at the shoulders, . . 4
7. SHOULDERS.—Shoulders broad and full, smoothly joining the neck with the back, ... 5
8. BREAST.—Breast wide, deep, and projecting well forward, the forelegs standing wide apart, ... 5
9. BACK AND LOIN.—Back and loin broad and straight from shoulders to rump, ... 7
10. RIBS.—Ribs well arched, extending far backward, the last projecting more than the others, ... 6
11. RUMP.—Rump broad, square and full, with tail well set up, 6
12. HIPS.—Hips wide, with little space between them and last ribs, ... 6
13. THIGHS.—Thighs full and well let down in twist, the legs standing well apart, ... 6
14. LIMBS.—Limbs short and fine in bone, and in color to agree with the face, ... 3
15. FORELEGS.—Forelegs well wooled and carrying mutton to the knees, but free from meat below, ... 2
16. HINDLEGS.—Well filled with mutton and wooled to the hocks, neat and clean below, ... 2
17. BELLY AND FLANK.—Belly straight and well covered with wool, the flank extending so as to form a line parallel with the back or top line, ... 5
18. FLEECE.—Fleece compact, the whole body well covered with moderately long and close wool, white in color, carrying some yolk, ... 12
19. FORM.—Form throughout smooth and symmetrical, with no coarseness in any part, ... 9
20. GENERAL APPEARANCE.—General appearance spirited and attractive, with a determined look, a proud and firm step, indicating constitutional vigor and thorough breeding, . 8

PERFECTION, - - - - 100

SUFFOLK SHEEP.

Standard of Excellence for Suffolk Sheep, as adopted by the American Suffolk Flock Registry Association, George W. Franklin, Secretary, Atlantic, Iowa.

SCALE OF POINTS FOR SUFFOLK SHEEP.

POINTS.	COUNTS
1. General Appearance,	7
2. General Form,	15
3. Head,	15
4. Neck,	5
5. Fore-quarters,	15
6. Barrel,	10
7. Hind-quarters,	15
8. Feet and Legs,	8
9. Fleece,	10
PERFECTION,	100

DETAILED DESCRIPTION.

POINTS.   COUNTS.

1. GENERAL APPEARANCE.—Pleasing outline; good carriage, and symmetry of development, ... 7
2. GENERAL FORM.—Large in size; inclined to be long in body; medium strength of bone; somewhat cylindrical in shape, and straight above, below and in the rear, ... 15
3. HEAD.—Medium in size, inclining to be long and covered with fine, short, glossy, black hair to the junction with the neck; a small quantity of clean, white wool on the forehead is not objected to; muzzle moderately fine, especially in ewes; eyes bright and full; ears of medium length and fineness, ... 15
4. NECK.—Moderately long and well set, and blending well with the body with some crest in the lambs, . . . 5
5. FORE-QUARTERS.—Well developed; breast wide, deep and full; brisket, broad; chest, capacious, with good heart girth; shoulders broad, oblique and well filled in the neck, vein and crops; withers broad; arm, well developed, . . 15
6. BARREL.—Roomy; back, straight, broad and well fleshed

throughout its entire length; ribs, well sprung and moderately deep; fore and hind flanks, full and deep, . . 10
7. HIND-QUARTERS.—Long, deep and full; tail, broad and well set up; buttock, broad; twist full; thigh, broad and full 15
8. FEET AND LEGS.—Straight, of medium length with flat bone; bare of wool below the knee and hock, glossy black in color and set well apart, ... 8
9. FLEECE.—Moderately short, with close, fine, lustrous fiber, and without tendency to mat or felt together, or to shade off into dark or gray wool or hair, especially about the neck and tail. The fleece should cover the whole body except the head and the legs below the knee and hock; and the skin underneath it should be fair, soft and of a pink color, ... 10

PERFECTION, - - - - 100

ANGORA GOATS.

In reply to a request, Mr. C. P. Bailey, of San Jose, California, the principal breeder of Angora Goats in the United States, sent the editor the following scale of points and detailed description, copied from the minutes of a meeting of the Angora Goat Breeders' Association, held September 22, 1887:

SCALE OF POINTS FOR ANGORA GOATS.

POINTS. COUNTS.

FLEECE—Thirty-one Points.

1. Fineness, ... 9
2. Weight, ... 8
3. Evenness, ... 6
4. Shape, ... 4
5. Lustre, ... 4

BODY—Sixteen Points.

6. Constitution, ... 6
7. Symmetry of Shape, ... 5
8. Weight, ... 5

EARS—Three Points.

9. Ear Lock, ... 2
10. Lop Ears, ... 1

Perfection, - - - - 50

DETAILED DESCRIPTION OF ANGORA GOATS.

A perfect goat when in full fleece should appear like a parallelogram. The body should be full and long and of straight build. It should be densely and evenly covered with long, lustrous, fine, curly hair, appearing from a distance as if it had been trimmed off below the body. The chest and shoulders, especially with the males, should be broad and strong, and legs straight and chunky; the head clear cut and trim, not coarse like that of a common goat. The horns of the buck are long and strong, inclined toward the back, and of spiral like shape. The horns of the does, short and thin, and curved backwards.

In Mr. Bailey's descriptive circular it is stated that the average fleece of pure-bred goats is from four to six pounds, but frequently eight and ten pounds have been obtained from choice, well-kept animals.

SWINE.

98 NOMENCLATURE FOR HOG.

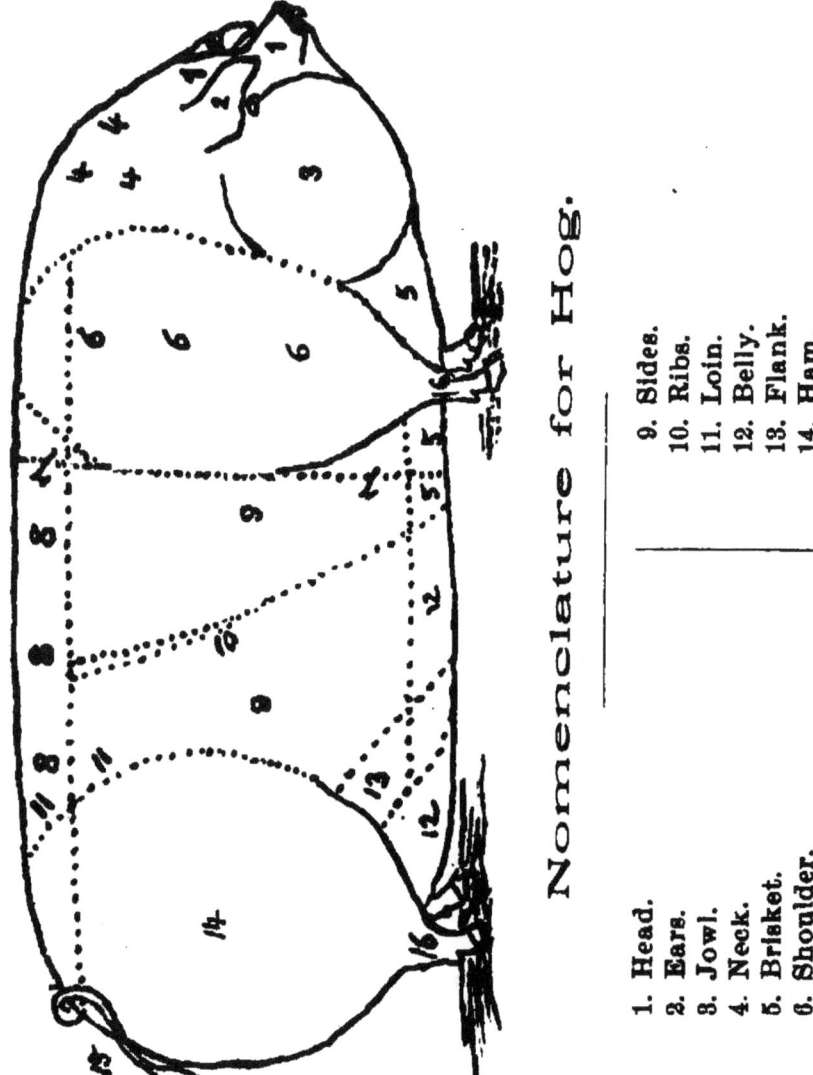

Nomenclature for Hog.

1. Head.
2. Ears.
3. Jowl.
4. Neck.
5. Brisket.
6. Shoulder.
7. Girth Around Heart.
8. Back.
9. Sides.
10. Ribs.
11. Loin.
12. Belly.
13. Flank.
14. Ham.
15. Tail.
16. Legs.

BERKSHIRE SWINE.

Standard of Excellence for Berkshire Swine, as adopted by The American Berkshire Association, Charles F. Mills, Secretary, Springfield, Illinois.

SCALE OF POINTS FOR BERKSHIRE SWINE.

POINTS.	COUNTS.
1. Color,	4
2. Face and Snout,	7
3. Eye,	2
4. Ear,	4
5. Jowl,	4
6. Neck,	4
7. Hair,	3
8. Skin,	4
9. Shoulder,	7
10. Back,	8
11. Side,	6
12. Flank,	5
13. Loin,	9
14. Ham,	10
15. Tail,	2
16. Legs,	5
17. Symmetry,	6
18. Condition,	5
19. Style,	5
PERFECTION,	100

DETAILED DESCRIPTION.

POINTS.	COUNTS.
1. COLOR—Black, with white on feet, face, tip of tail, and an occasional splash on the arm,	4
2. FACE AND SNOUT—Short; the former fine and well dished, and broad between the eyes,	7
3. EYE—Very clear, rather large, dark hazel or gray,	2
4. EAR—Generally almost erect, but sometimes inclined forward with advancing age; medium size; thin and soft,	4
5. JOWL—Full and heavy, running well back on neck,	4
6. NECK—Short and broad on top,	4
7. HAIR—Fine and soft; medium thickness,	3

8. SKIN—Smooth and pliable, ... 4
9. SHOULDER—Thick and even, broad on top, and deep through chest, ... 7
10. BACK—Broad, short and straight; ribs well sprung, coupling close to hips, ... 8
11. SIDE—Deep and well let down; straight on bottom lines, . 6
12. FLANK—Well back, and low down on leg, making nearly a straight line with lower part of side, ... 5
13. LOIN—Full and wide, ... 9
14. HAM—Deep and thick, extending well up on back, and holding thickness well down to hock, ... 10
15. TAIL—Well set up on back; tapering and not coarse, . . 2
16. LEGS—Short, straight and strong; set wide apart, with hoofs erect, and capable of holding good weight, . . . 5
17. SYMMETRY—Well proportioned throughout, depending largely on condition, ... 6
18. CONDITION—In a good, healthy growing state; not over fed 5
19. STYLE—Attractive, spirited, indicative of thorough breeding and constitutional vigor, ... 5

PERFECTION, - - - - 100

Standard of Excellence for Berkshire Swine, as adopted by the National Association of Expert Judges on Swine, W. M. Lambing, Secretary, West Liberty, Iowa, and by The National Berkshire Record Association, E. K. Morris, Secretary, 467 South Illinois Street, Indianapolis, Indiana.

SCALE OF POINTS FOR BERKSHIRE SWINE.

POINTS.	COUNTS
1. Head and Face,	4
2. Eyes,	2
3. Ears,	2
4. Neck,	2
5. Jowl,	2
6. Shoulder,	6
7. Chest,	12
8. Back,	15
9. Sides and Ribs,	8
10. Belly and Flank,	6
11. Ham and Rump,	10
12. Legs and Feet,	10
13. Tail,	1

14. Coat, ... 2
15. Color, ... 2
16. Size, ... 5
17. Action and Style, ... 4
18. Condition, ... 4
19. Disposition, ... 8

PERFECTION, - - - - 100

DISQUALIFICATIONS.

FORM: Very large and heavy or drooping ears; small cramped chest; crease back of shoulders and over the back so as to cause a depression in back easily noticed; deformed or crooked legs; feet broken down, so that the animal walks on pastern joints.

SIZE: Overgrown, gangling, narrow, contracted or not two-thirds large enough for age.

CONDITION: Barrenness; deformed; seriously diseased; total blindness from any cause.

SCORE: Less than sixty points.

PEDIGREE: Not eligible to record.

DETAILED DESCRIPTION.

POINTS. COUNTS.

1. HEAD AND FACE—Head short; broad; coming well forward at poll; face short and fine and well dished; broad between the eyes, tapering from eyes to point of nose, surface even and regular, ... 4

    Objections—Head long and narrow; coarse, forehead low and narrow; jaws narrow or contracted, lower jaws extending beyond upper; face long; straight between eyes; nose coarse, thick, or crooked, or ridgy.

2. EYES—Very clear; rather large, dark-hazel or gray, . . 2

    Objections—Small, dull, bloodshot, deepset or obscure, vision impaired by wrinkles, fat or other cause.

3. EARS—Generally almost erect, but sometimes inclined forward with advancing age; medium size, thin and soft, . 2

    Objections—Large, coarse, thick, round or drooping; long or large knuck; difference in form, size or position one with the other; animal not being able to control their position.

4. NECK—Full, deep, short, and slightly arched; broad on top, well connected with shoulder, ... 2

    Objections—Long, flat; lacking in fullness and depth.

5. JOWL—Full, firm and neat; carrying fullness back to shoulder and brisket, ... 2

Objections—Light, flabby, thin, tucked up or wrinkled.

6. SHOULDER—Broad, deep and full, not extended above line of back and being as wide on top as back, carrying size down to line of belly, and having lateral width, . . 6

    Objections—Lacking in depth or width, thick beyond the line of sides and hams or extending above line of back; heavy shields on hogs under eighteen months of age.

7. CHEST—Large, wide, deep and roomy; full girth; breast bone curving well forward; extending back on level; not tucked up; broad between forelegs, ... 12

    Objections—Flat; narrow at top or bottom; small girth; lacking depth or fullness; breast bone crooked or tucked up.

8. BACK—Broad and straight, carrying same width from shoulder to ham, surface even and smooth without creases or projections and not too long, ... 15

    Objections—Narrow, swayed or hollow, dropping below a straight line.

9. SIDES AND RIBS—Sides full, smooth, firm and deep; carrying size down to belly and evenly from ham to shoulder; ribs long, strong, well sprung at top and bottom, . . 8

    Objections—Flat, thin, flabby; not as full at bottom as top. Ribs weak, not well sprung at top or bottom.

10. BELLY AND FLANK—Wide, full, and straight on bottom line, ... 6

    Objections—Belly narrow and sagging. Flank thin and tucked up.

11. HAM AND RUMP—Hams broad, full and long; the lower front part of ham should be full and stifle well covered with flesh; coming well down to hock. Rump should have a rounding slope from loin to root of tail; same width as back and filling out on each side and above the tail, . . 10

    Objections—Ham narrow, short, thin; not projecting beyond and coming down on hock; cut up too high in crotch. Rump flat, narrow and too steep.

12. LEGS AND FEET—Legs short, straight and strong; set wide apart with hoofs erect and capable of holding good weight, ... 10

    Objections—Legs long, slim, coarse, crooked; muscles light, pastern long, slim or flat, feet long or sprawling.

13. TAIL—Set well up, fine, tapering and neatly curled, . . 1
   Objections—Coarse and straight ; too low.
14. COAT—Fine, straight, smooth ; laying close to and covering
   the body well ; not clipped ; evenly distributed over body, 2
   Objections—Hair coarse, harsh, wavy or curly ; not evenly
   distributed over body ; swirls or clipped.
15. COLOR—Black, with white on feet, face, tip of tail and an
   occasional splash on arm, ... 2
   Objections—Solid black or black points, or white spots on
   body.
16. SIZE—Large for age. Boar two years and over not less than
   450 pounds ; sows same age 400 pounds. Boars eighteen
   months, 350 pounds ; sows same age, 325 pounds. Boars
   twelve months, 300 pounds ; sows same age, 275 pounds.
   Boars and sows six months, 150 pounds, . . . 5
   Objections—Under weight; coarse ; not in good form to fatten.
17. ACTION and STYLE—Action vigorous. Style, graceful and
   attractive, ... 4
   Objections—Dull, sluggish and clumsy.
18. CONDITION—Healthy ; skin clear of scurf, scales or sores,
   soft and mellow to the touch ; flesh fine, evenly laid on
   and free from lumps ; hair soft and lying close to body ;
   good feeding qualities, ... 4
   Objections—Unhealthy ; skin scaly, scabby or harsh, flabbiness or lumpy flesh ; too much fat for breeding. Hair
   harsh, dry and standing up from body ; poor feeders ;
   deafness, partial or total.
19. DISPOSITION—Quiet and gentle and easy to handle, . . 3
   Objections—Cross, restless, vicious or wild.

     PERFECTION, - - - - 100

CHESHIRE SWINE.

Standard of Excellence for Cheshire Swine, as adopted by the Cheshire Swine Breeders' Association, R. D. Button, Secretary, Cottons, N. Y.

SCALE OF POINTS FOR CHESHIRE SWINE.

POINTS.	COUNTS.
1. Head,	8
2. Face,	8
3. Jowl,	8
4. Ears,	5
5. Neck,	8
6. Shoulders,	6
7. Girth Around Heart,	8
8. Back,	10
9. Sides,	7
10. Flank,	3
11. Hams,	10
12. Legs,	10
13. Tail,	3
14. Hair,	3
15. Color,	2
16. Skin,	3
17. Symmetry,	8
PERFECTION,	100

DETAILED DESCRIPTION.

POINTS.	COUNTS.
1. HEAD—Short to medium in length short in proportion to length of body,	8
2. FACE—Somewhat dished and wide between the eyes,	8
3. JOWL—Medium in fullness,	3
4. EARS—Small, fine, erect, and in old animals slightly pointing forward,	5
5. NECK—Short and broad,	8
6. SHOULDERS—Broad, full and deep,	6
7. GIRTH AROUND HEART—	8
8. BACK—Long, broad and straight nearly to root of tail,	10
9. SIDES—Deep and full; nearly straight on bottom line,	7
10. FLANK—Well back and low down making flank girth nearly	

equal to heart girth, ... 8
11. HAMS—Broad and nearly straight with back and running well down towards hock, ... 10
12. LEGS—Small and slim, set well apart, supporting body well on toes, ... 10
13. TAIL—Small, slim and tapering, ... 3
14. HAIR—Fine, medium in thickness and quantity, . . 3
15. COLOR—White, any colored hairs to disqualify, . . 2
16. SKIN—Fine and pliable, small blue spots objectionable but allowab'e, ... 3
17. SYMMETRY—Animal well proportioned, handsome and stylish, and when grown and well fattened should dress from 400 to 600 pounds, ... 8

    PERFECTION, - - . - 100

CHESTER WHITE SWINE.

Standard of Excellence for Chester White Swine, as adopted by the National Chester White Record Association, Thomas Sharpless, Secretary, West Chester, Pennsylvania.

SCALE OF POINTS FOR CHESTER WHITE SWINE.

POINTS.	COUNTS.
1. Color—White,	8
2. Head—Small, broad, and face slightly dished,	5
3. Ears—Fine and drooping,	2
4. Jowl—Neat and full,	2
5. Neck—Short, full and slightly arched,	3
6. Brisket—Full,	3
7. Shoulders—Broad and deep,	6
8. Girth Around the Heart—	10
9. Back—Straight and broad,	7
10. Sides—Deep and full,	6
11. Ribs—Well sprung,	7
12. Loin—Broad and strong,	7
13. Belly—Wide and straight,	4
14. Flank—Well let down,	3
15. Ham—Broad, full and deep,	10
16. Tail—Tapering, not coarse,	2
17. Limbs—Strong, straight and tapering,	7
18. Coat—Thick and soft,	3
19. Action—Prompt, easy and graceful,	5
20. Symmetry—Adaptation of the several parts to each other,	5
PERFECTION,	100

DETAILED DESCRIPTION.

POINTS.   COUNTS.

1. COLOR—White, ... 8

   Objections—Blue spots on skin shall argue impurity; and black spots in hair disqualify them and their offspring.

2. HEAD—Short, broad between the eyes, and nicely tapering from eyes to point of nose; face slightly dished; cheeks full, ... 5

   Objections—Head coarse; face long and narrow, and too much dished; snout coarse and thick.

3. EARS—Drooping, fine and silky, pointing forward and a

little outward; well proportioned to size of body, . . 2
   Objections—Too large and coarse; thick, lopping and lying too near the face; stiff, erect or too round.
4. JOWL—Full, firm, neat and carrying fullness well back to shoulders and brisket, ... 2
   Objections—Flabby, light, too thick in cheek, tucking up under the throat.
5. NECK—Full, deep, short and slightly arched, . . 3
   Objections—Long, flat or narrow.
6. BRISKET—Full, well let down, and well joined to jowl and in a line with belly, ... 3
   Objections—Narrow or tucked up.
7. SHOULDERS—Broad, deep, thickness in proportion to the sides and hams, and full and even on top, ... 6
   Objections—Lacking in depth or width, thick beyond the line of side and ham, or blade too prominent.
8. GIRTH AROUND THE HEART—Full back of the shoulders, ribs extending well down; wide and full back of fore legs, 10
   Objections—Less than flank measure, or length of body from top of head to root of tail, or creased back of shoulders.
9. BACK—Broad, straight or slightly arched, carrying width well back to hams, and of medium length, . . . 7
   Objections—Narrow, creasing back of shoulder, narrow across the loins, swayed, too long or sun-fish shape.
10. SIDES—Full, deep, carrying size well down and back, . . 6
    Objections—Too round or flat, shallow or thin at flank.
11. RIBS—Well sprung and long, carrying fullness and depth well back, ... 7
    Objections—Too flat, or curve too short.
12. LOIN—Broad, strong and full, ... 7
    Objections—Narrow and weak.
13. BELLY—Wide and straight, ... 4
    Objections—Sagging; narrow.
14. FLANK—Well let down and full, ... 3
    Objections—Thin, tucked in or cut up too high.
15. HAM—Full, broad, deep, holding width and coming down well over hock, ... 10
    Objections—Narrow, short, too steep at rump, or cut up too high in crotch.
16. TAIL—Well set on, small, smooth and well tapered, . . 2

Objections—Coarse, too large or too prominent at root.

17. LIMBS—Medium length, set well apart and well tapered. Bone firm and flinty; muscles full above knee and hock; pastern and foot both short. 7
Objections—Long, slim, coarse, crooked, muscles light, pastern long, slim or flat; foot long or sprawling.

18. COAT—Fine, thick, and covering the body well, . . 3
Objections—Coarse, bristly, harsh, wiry.

19. ACTION—Easy, prompt and graceful, ... 5
Objections—Dull, sluggish and clumsy.

20. SYMMETRY—A harmonious combination of the foregoing Scale of Points, ... 5
Objections—Too much development in some points and lacking in others.

    PERFECTION, - - - - 100

Standard of Excellence for Chester White Swine, as adopted by The Chester White Record Association, W. H. Morris, Secretary, Indianapolis, Indiana; also by The National Association of Expert Judges on Swine, W. M. Lambing, Secretary, West Liberty, Iowa.

SCALE OF POINTS FOR CHESTER WHITE SWINE.

POINTS.	COUNTS
1. Head and face,	4
2. Eyes,	2
3. Ears,	2
4. Neck,	2
5. Jowl,	2
6. Shoulders,	6
7. Chest,	12
8. Back and Loin,	15
9. Sides and Ribs,	8
10. Belly and Flank,	6
11. Ham and Rump,	10
12. Legs and Feet,	10
13. Tail,	1
14. Coat,	2
15. Color,	2

16. Size, ... 5
17. Action and Style, ... 4
18. Condition, ... 4
19. Disposition, ... 3

                    PERFECTION, - - - 100

DISQUALIFICATIONS.

FORM: Upright ears; small, cramped chest; crease around back of shoulders and over the back, causing a depression easily noticed; feet broken down, causing the animal to walk on joints; deformed or badly crooked legs.

SIZE: Chuffy or not two-thirds large enough for age.

CONDITION: Squabby fat; deformed, seriously diseased; barrenness; total blindness.

SCORE: Less than sixty points.

PEDIGREE: Not eligible to record.

COLOR: Black or sandy spots in hair.

DETAILED DESCRIPTION.

POINTS.   COUNTS.

1. HEAD and FACE—Head short and wide; cheeks neat but not too full; jaws broad and strong; forehead medium, high and wide; face short and smooth; wide between the eyes; nose neat and tapering and slightly dished, . . 4

   Objections—Head long, narrow and coarse; forehead low and narrow; jaws contracted and weak; face long, narrow and straight; nose coarse, clumsy or dished like a Berkshire.

2. EYES—Large, bright, clear and free from wrinkles or fat surroundings, ... 2

   Objections—Small, deep or obscure; vision impaired in any way.

3. EARS—Medium size; not too thick; soft; attached to the head so as not to look clumsy; pointing forward and slightly outward; fully under the control of the animal and drooping so as to give a graceful appearance. . . 2

   Objections—Large; upright; coarse; thick; round; too small; drooping too close to the face; animal not being able to control them.

4. NECK—Wide; deep; short and nicely arched, . . . 2

   Objections—Long; narrow; thin; flat on top; tucked up; not extending down to breast bone.

5. JOWL—Full; smooth; neat and firm; carrying fulness back to shoulder and brisket when the head is carried up level. 2
   Objections—Light; too large and flabby; rough and deeply wrinkled; not carrying fullness back to shoulder and brisket.

6. SHOULDER—Broad, deep and full, extending in a straight line with the side, and carrying size down to line of belly. 6
   Objections—Narrow at top or bottom, not full nor same depth as body; extending above line of back; shields on boars too coarse and prominent.

7. CHEST—Large; deep and roomy so as not to cramp vital organs; full in girth around the heart; the breast bone extending forward so as to show slightly in front of legs, and let down so as to be even with line of belly, showing a width of not less than 7 inches between forelegs of a full grown hog, ... 12
   Objections—Narrow; pinched; heart girth less than flank girth; too far let down between forelegs; breast bone crooked or too short.

8. BACK and LOIN—Back broad on top; straight or slightly arched; uniform width; smooth; free from lumps or rolls; shorter than lower belly line; same height and width at shoulder as at ham; loin wide and full, . . 15
   Objections—Back narrow; creased back of shoulders; sunfished shape; humped; swayed; too long or lumpy rolls; uneven in width; loin narrow, depressed or humped.

9. SIDES and RIBS—Sides full; smooth; deep; carrying size down to belly; even with line of ham and shoulder; ribs long; well sprung at top and bottom, giving hog a square form, ... 8
   Objections—Flat; thin; flabby; compressed at bottom; shrunken at shoulder and ham; uneven surface; ribs flat and too short.

10. BELLY and FLANK—Same width as back; full, making a straight line and dropping as low at flank as at bottom of chest; line of lower edge running parallel with sides; flank full and even with body, ... 6
    Objections—Belly narrow; pinched; sagging or flabby; flank thin, tucked up or drawn in.

11. HAM and RUMP—Ham broad; full; long; wide and deep; admitting of no swells; buttock full; neat and clean, thus avoiding flabbiness; stifle well covered with flesh, nicely tapering towards the hock; rump should have a slightly rounding shape from loin to root of tail; same width as back, making an even line with sides, . . . 10

Objections—Ham narrow; short; not filled out to stifle; too much cut up in crotch or twist; not coming down to hock; buttocks flabby; rump flat, narrow, too long, too steep, sharp or peaked at root of tail.

12. LEGS and FEET—Legs short; straight; set well apart and squarely under body; bone of good size; firm; well muscled; wide above knee and hock; below knee and hock round and tapering, enabling the animal to carry its weight with ease; pasterns short and nearly upright feet short, firm, tough and free from defects, . . . 10

Objections—Legs too short; long; slim; crooked; too coarse; too close together; weak muscles above hock and knee; bone large and coarse, without taper; pasterns long; crooked, slim like a deer's; hoofs long, slim; weak; toes spreading, crooked or turned up.

13. TAIL—Small; smooth; tapering, well set on; root slightly covered with flesh; carried in a curl, ... 1

Objections—Coarse; long; clumsy; set too high or too low, hanging like a rope.

14. COAT—Fine; straight or wavy; evenly distributed and covering the body well; nicely clipped coats no objection, . 2

Objections—Bristles; hair coarse; thin; standing up; not evenly distributed over all the body except the belly.

15. COLOR—White (blue spots or black specks in skin shall not argue impurity of blood), ... 2

Objections—Color any other than white.

16. SIZE—Large for age and condition; boars two years old and over, if in good flesh, should weigh not less than 500 lbs. Sows same age and condition, not less than 450 lbs. Boars 18 months old in good flesh should weigh not less than 400 pounds. Sows, 350. Boars twelve months old not less than 300 pounds; sows 300. Boars and sows 6 months old, not less than 150 lbs each, and other ages in proportion, . 5

Objections—Overgrown; coarse; uncouth; hard to fatten.

17. ACTION and STYLE—Action easy and graceful; style attractive; high carriage; in males testicles should be readily seen; same size and carriage, . . . , . . 4
    Objections—Sluggish; awkward low carriage; wabbling walk; in males testicles not easily seen; not of same size or carriage, or only one showing.
18. CONDITION—Healthy; skin clear and bright; free from scurf and sores; flesh fine and mellow to the touch; evenly laid on and free from lumps; good feeding qualities, . 4
    Objections—Unhealthy; skin scaly, scabby or harsh; flesh lumpy or flabby; hair harsh, dry and standing up from body; poor feeders; total deafness.
19. DISPOSITION—Quiet; gentle and easily handled; with ambition enough to look out for themselves if neglected, . 3
    Objections—Cross; restless; vicious or wild; no ambition.

        PERFECTION, - - - - 100

DUROC-JERSEY SWINE.

Standard of Excellence for Duroc-Jersey Swine, as adopted by the American Duroc-Jersey Swine Breeders' Association, S. E. Morton, Secretary, Camden, Ohio; by the National Duroc-Jersey Record Association, R. J. Evans, Secretary, El Paso, Illinois; and by the National Association of Expert Judges On Swine W. M. Lambing, Secretary, West Liberty, Iowa.

SCALE OF POINTS FOR DUROC-JERSEY SWINE.

POINTS.	COUNTS.
1. Head and Face,	4
2. Eyes,	2
3. Ears,	2
4. Neck,	2
5. Jowl,	2
6. Shoulders,	6
7. Chest,	12
8. Back and Loin,	15
9. Sides and Ribs,	8
10. Belly and Flank,	6
11. Ham and Rump,	10
12. Legs and Feet.	10
13. Tail,	1
14. Coat.	2
15. Color,	2
16. Size,	5
17. Action and Style.	4
18. Condition,	4
19. Disposition,	3
PERFECTION,	100

DISQUALIFICATIONS.

FORM: Ears standing erect; small cramped chest and crease back of shoulders and over back so as to cause a depression in the back easily noticed; seriously deformed legs, or badly broken down feet.

SIZE: Very small, or not two-thirds large enough as given by the standard.

SCORE: Less than fifty points.

PEDIGREE: Not eligible to record.

DETAILED DESCRIPTION.

POINTS. COUNTS

1. HEAD AND FACE—Head small in proportion to size of body; wide between eyes; face nicely dished (about half way between a Poland-China and a Berkshire) and tapering well down to the nose; surface smooth and even, . . . 4

    Objections—Large and coarse; narrow between the eyes; face straight; crooked nose, or too much dished.

2. EYES—Lively, bright and prominent, ... 2

    Objections—Dull, weak and obscure.

3. EARS—Medium; moderately thin; pointing forward, downward and slightly outward, carrying a slight curve, attached to head very neatly, ... 2

    Objections—Very large; nearly round; too thick; swinging or flabby; not of same size; different position and not under control of animal.

4. NECK—Short, thick, and very deep and slightly arching, . 2

    Objections—Long, shallow and thin.

5. JOWL—Broad, full and neat; carrying fullness back to point of shoulders and on a line with breast bone, . . . 2

    Objections—Too large, loose and flabby, small, thin and wedging.

6. SHOULDERS—Moderately broad; very deep and full; carrying thickness well down and not extending above line of back, ... 6

    Objections—Small; thin; shallow; extending above line of back. Boars under one year old heavily shielded.

7. CHEST—Large; very deep; filled full behind shoulders; breast-bone extending well forward so as to be readily seen, ... 12

    Objections—Flat, shallow, or not extending well down between forelegs.

8. BACK AND LOIN—Back medium in breadth; straight or slightly arching; carrying even width from shoulder to ham; surface even and smooth, ... 15

    Objections—Narrow; creased behind shoulders; swayed or humped backed.

9. SIDES AND RIBS—Sides very deep; medium in length; level between shoulders and hams and carrying out full down to line of belly. Ribs long, strong and sprung in proportion to width of shoulders and hams, ... 8

Objections—Flabby, creased, shallow and not carrying proper width from top to bottom.

10. BELLY AND FLANK—Straight and full and carrying well out to line of sides. Flank well down to lower line of sides, . . 6
    Objections—Narrow; tucked up or drawn in; sagging or flabby.

11. HAMS AND RUMP—Broad, full and well let down to the hock; buttock full and coming nearly down and filling full between hocks. Rump should have a round slope, from loin to root of tail; same width as back and well filled out around tail, ... 10
    Objections—Ham narrow; short; thin; not projecting well down to hock; cut up too high in crotch. Rump narrow; flat or peaked at root of tail; too steep.

12. LEGS AND FEET—Medium size and length; straight; nicely tapered; wide apart and well set under the body; pasterns short and strong. Feet short, firm and tough, . . 10
    Objections—Legs extremely long, or very short; slim; coarse; crooked; legs as large below knee and hock as above; set too close together; hocks turned in or out of straight line. Feet—hoofs long, slim and weak; toes spreading or crooked.

13. TAIL—Medium; large at base and nicely tapering and rather bushy at end, ... 1
    Objections—Extremely heavy; too long and ropy.

14. COAT—Moderately thick and fine; straight, smooth and covering the body well, ... 2
    Objections—Too many bristles; hair coarse, harsh and rough; wavy or curly; swirls, or not evenly laid over the body.

15. COLOR—Cherry red without other admixtures, . . . 2
    Objections—Very dark red or shading brown; very pale or light red; black spots over the body; black flecks on belly and legs not desired but admissable.

16. SIZE—Large for age and condition. Boars two years old and over should weigh 600 pounds; sows same age and condition, 500 pounds. Boars, eighteen months, 475 pounds; sows, 400 pounds. Boars, twelve months, 350 pounds; sows, 300 pounds; Boar and sow pigs six months, 150 pounds. These figures are for animals in a fair show

condition. 5
    Objections—Rough and coarse and lacking in feeding qualities.

17. ACTION AND STYLE—Action vigorous and animated. Style free and easy. 4
    Objections—Dull or stupid; awkward and wabbling. In boars testicles not easily seen nor of same size or carriage; too large or only one showing.

18. CONDITION—Healthy; skin free from any scurf, scales, sores and mange; flesh evenly laid over the entire body and free from any lumps, ... 4
    Objections—Unhealthy; scurfy; scaley; sores; mange; too fat for breeding purposes; hair harsh and standing up; poor feeders.

19. DISPOSITION—Very quiet and gentle; easily handled or driven, ... 3
    Objections—Wild, vicious or stubborn.

                          PERFECTION, - - - - 100

ESSEX SWINE.

Standard of Excellence for Essex Swine, as adopted by the American Essex Association. F. M. Srout, Secretary, McLean, Illinois.

SCALE OF POINTS FOR ESSEX SWINE.

POINTS.	COUNTS.
1. Color—Black,	2
2. Head—Small, broad and face dished,	8
3. Ears—Fine, erect, slightly drooping with age,	2
4. Jowl—Full and neat,	1
5. Neck—Short, full and slightly arched,	8
6. Shoulders—Broad and deep,	7
7. Girth around heart,	6
8. Back—Straight, broad and level,	12
9. Side—Deep and full,	6
10. Ribs—Well sprung,	7
11. Loin—Broad and strong,	12
12. Flank—Well let down,	2
13. Ham—Broad, full and deep,	12
14. Tail—Medium, fine, and curled,	2
15. Legs—Fine, straight and tapering,	3
16. Feet—Small,	3
17. Hair—Fine and silky, free from bristles,	3
18. Action—Easy and graceful,	4
19. Symmetry—Adaption of the several parts to each other,	10
PERFECTION,	100

POLAND-CHINA SWINE.

Standard of Excellence for Poland-China Swine, as adopted by the National Poland-China Breeders' Association, E. C. Rouse, Fecretary, Albion, Michigan. Also by the Ohio Poland-China Record Company, Carl Freigau, Secretary, Dayton, Ohio. Also by the American Poland-China Record Company, W. M. McFadden, Secretary, West Liberty, Iowa. Also by the Standard Poland-China Record Association, George F. Woodworth, Secretary, Maryville, Missouri. Also by the Northwestern Poland-China Record Association J. B. Besack, Secretary, Washington, Kansas.

SCALE OF POINTS FOR POLAND-CHINA SWINE.

POINTS.   COUNTS

1. Color—Dark spotted or black,   8
2. Head—Small, broad, face slightly dished,   5
3. Ears—Fine and drooping,   2
4. Jowl—Neat and full,   2
5. Neck—Short, full, slightly arched,   3
6. Brisket—Full,   3
7. Shoulder—Broad and deep,   6
8. Girth around heart,   10
9. Back—Straight and broad,   7
10. Loin—Broad and strong,   7
11. Sides—Deep and full,   6
12. Ribs—Well sprung,   7
13. Belly—Wide and straight,   4
14. Flank—Well let down,   3
15. Ham—Broad, full and deep,   10
16. Tail—Tapering and not coarse,   2
17. Limbs—Strong, straight and tapering,   7
18. Coat—Thick and soft,   3
19. Action—Prompt, easy and graceful,   5
20. Symmetry—Adaptation of the several points to each other,   5

                PERFECTION,   100

DISQUALIFICATIONS.

CONDITION: Excessive fatness; barren; deformed; unsound or diseased; ridgling or one-seeded. More than one-half white or sandy.

SCORE: A score of less than sixty of the standard.

PEDIGREE: Lack of eligibility to record.

DETAILED DESCRIPTION.

POINTS. COUNTS.

1. COLOR—Black or dark spotted with white points. (Sandy spots and speckled color shall not argue impurity of blood, but are not desirable). 3

   Objections—Solid black or with more sandy or white than black hairs over body.

2. HEAD—Short, broad between the eyes and nicely tapering from eyes to point of nose; face slightly dished; cheeks full, ... 5

   Objections—Head coarse, long and narrow; face too much dished; snout coarse and thick.

3. EARS—Drooping, fine and silky; pointing forward and a little outward; well proportioned to size of body, . . 2

   Objections—Too large and coarse; thick, lopping; lying too near the face; stiff, erect or too round.

4. JOWL—Full, firm and neat; carrying fullness well back to shoulder and brisket, ... 2

   Objections—Flabby; light; too thin in cheeks; tucking up under the neck.

5. NECK—Full, deep, short and slightly arched, . . . 3

   Objections—Long; flat; lacking in fullness or depth.

6. BRISKET—Full; well let down, extending well forward and on line of the belly, ... 3

   Objections—Narrow or tucked up.

7. SHOULDERS—Broad, deep; thickness in proportion to the sides and hams; full and even on top, ... 6

   Objections—Lacking in depth or width, thick beyond the line of the sides and hams; blade too prominent.

8. GIRTH AROUND HEART—Full back of shoulders; ribs extending well down, wide and full back of forelegs, . . 10

   Objections—Less than flank measure or length of body from top of head to root of tail, or creased back of shoulders,

9. BACK—Broad, straight or slightly arched carrying width well back to hams and of medium length, . . . 7

   Objections—Narrow; creasing back of shoulder; narrow across the loins; swayed; too long; sunfish shape.

10. LOIN—Broad, strong and full, ... 7

    Objections—Narrow; weak.

11. SIDES—Full, deep, carrying size well down and back, . . 6

    Objections—Too round or flat; shallow or thin at the flank,

120   POLAND-CHINA SWINE.

12. RIBS—Well sprung and long, carrying fullness and depth well back, ... 7
    Objections—Too flat; curve of rib too short.
13. BELLY—Wide and straight, ... 4
    Objections—Sagging; narrow.
14. FLANK—Well let down and full, ... 8
    Objections—Thin, tucked in, cut up too high.
15. HAM—Full, broad, deep, holding width and coming down well over hock, ... 10
    Objections—Narrow, short, too deep at the rump and cut up too high in crotch.
16. TAIL—Well set on, small, smooth and tapering, . . . 2
    Objections—Coarse, large, too prominent at the root.
17. LIMBS—Medium length, well set apart and well tapered, bone firm and flinty, not coarse, muscles full above knee and hock, pastern short, foot short, ... 7
    Objections—Long, slim, coarse, crooked, muscles light, pastern long, slim or flat, feet long or sprawling.
18. COAT—Fine, thick and covering the body well, . . . 3
    Objections—Coarse, bristly, harsh, and wiry.
19. ACTION—Easy, prompt, fine and graceful, ... 5
    Objections—Dull, sluggish, clumsy.
20. SYMMETRY—A harmonious combination of the foregoing scale of points, ... 5
    Objections—Too much developed in some points and lacking in others.

        PERFECTION, - - - - 100

SERIOUS OBJECTIONS.

FORM—Small growth; upright ears; small, cramped chest; crease back of the shoulders, so as to be readily seen; deformed and badly crooked legs; feet broken down so that the animal walks on pastern joint and dew claws.

Standard of Excellence for Poland-China Swine, as adopted by the Central Poland China Record Association, W. H. Morris, Secretary, Indianapolis, Indiana: also by the National Association of Expert Judges on Swine, W. M. Lambing, Secretary, West Liberty, Iowa.

SCALE OF POINTS FOR POLAND-CHINA SWINE.

POINTS.   COUNTS.
1. Head and Face, ... 4

POLAND-CHINA SWINE.

2.	Eyes,	2
3.	Ears,	2
4.	Neck,	2
5.	Jowl,	2
6.	Shoulders,	6
7.	Chest,	12
8.	Back and Loin,	15
9.	Sides and Ribs,	8
10.	Belly and Flank,	6
11.	Hair and Rump,	10
12.	Legs and Feet,	10
13.	Tail,	1
14.	Coat,	2
15.	Color,	2
16.	Size,	5
17.	Action and Style,	4
18.	Condition,	4
19.	Disposition,	3
	PERFECTION,	100

DISQUALIFICATIONS.

FORM: Upright ears; small cramped chest, crease back of shoulders and over the back so as to cause a depression in back easily noticed; deformed or badly crooked legs; feet broken down, so that the animal walks on pastern joints.

SIZE: China build, or not two-thirds large enough for age.

CONDITION: Excessive fatness; barrenness; deformed; seriously diseased; total blindness, caused by defective eyes, or by reason of fat or loose and wrinkled skin over the eyes.

SCORE: Less than sixty points.

PEDIGREE: Not eligible to record.

POINTS. **DETAILED DESCRIPTION.** COUNTS.

1. HEAD AND FACE—Head short and wide; cheeks full; jaws broad; forehead high and wide; face short; smooth; wide between eyes; tapering from eyes to point of nose and slightly dished; surface even and regular, ... 4
   Objections—Head long; narrow; coarse; forehead low and narrow or contracted; lower jaw extending beyond upper; face long, straight and narrow between eyes; nose coarse, thick or crooked, ridgy or dished as much as a Berkshire.

2. EYES—Large; prominent; bright; lively, clear, and free from wrinkled or fat surroundings, ... 2
   Objections—Small; dull; blood-shot; deep set or obscure; vision impaired by wrinkles, fat or other cause.

3. EARS—Small; thin; soft; silky; attached to the head by a

short and small knuck; tips pointing forward and slightly outward, and the forward half drooping gracefully; fully under control of animal; both of same size, position and shape, ... 2
   Objections—Large; straight; stiff; coarse; thick; round; long or large knuck; dropping close to face; swinging and flabby; difference in form, size or position.

4. NECK—Wide; deep; short, and nicely arched at top, from poll of head to shoulder, ... 2
   Objections—Long; narrow; thin; flat on top; not extending down to breast bone; tucked up.

5. JOWL—Full; broad; deep; smooth, and firm; carrying fullness back near to point of shoulders, and below line of lower jaw, so that lower line will be as low as breast bone when head is carried up level, ... 2
   Objections—Light; flabby; thin; wedge shaped; deeply wrinkled; not drooping below line of lower jaw, and not carrying fullness back to shoulder and brisket.

6. SHOULDERS—Broad; deep and full; not extending above line of back, and being as wide on top as back; carrying size down to line of belly, and having good lateral width, 6
   Objections—Narrow; not same depth as body; narrow at top or bottom or extending above line of back; less than body in breadth at top or bottom portions, or lacking in lateral width; shields on boars under eight months of age, or large, heavy shields, on hogs under eighteen months of age.

7. CHEST—Large; wide; deep; roomy, indicating plenty of room for vital organs, and making a large girth just back of shoulders; the breast bone extending forward so as to show slightly in front of legs, and extending in a straight line back to end of breast bone; showing a width of not less than six inches between forelegs in a large full grown hog, ... 12
   Objections—Flat; pinched; narrow at top or bottom; drawn or tucked underneath between forelegs or at either end of breast bone; breast bone crooked or not extending slightly in front of forelegs.

8. BACK AND LOIN—Broad; straight; or slightly arched; carrying same width from shoulder to ham; surface even;

smooth, free from lumps, creases or projections; not too long, but broad on top, indicating well sprung ribs; should not be higher at hip than at shoulder, and should fill out at junction with side, so that a straight-edge placed along top of side will touch all the way from point of shoulder to point of ham; should be shorter than lower belly line, 15

Objections—Narrow; creased back of shoulders; swayed or hollow; dropping below a straight line; humped or wrinkled; too long, or sunfish shaped; loin high, narrow, depressed, or humped up; surface lumpy, creased, ridgy or uneven; width at sides not as much as shoulder and ham.

9. SIDES AND RIBS—Sides full; smooth; firm and deep; carrying size down to belly and evenly from ham to shoulder; ribs long, strong, well sprung at top and bottom, . . 8

Objections—Flat; thin; flabby; pinched; not as full at bottom as top; drawn in at shoulders so as to produce a crease, or pinched and tucked up, and in, as it approaches the ham; lumpy or uneven surface; ribs flat or too short.

10 BELLY AND FLANK—Wide, straight and full, and dropping as low at flank as bottom of chest, back of foreleg, making a straight line from forelegs to hindlegs; flank full and out even with surrounding portions of body; the belly at that point dropping down on a line with lower line of chest; the loose skin connecting ham and belly, being on line even with bottom of side. 6

Objections—Belly narrow; pinched; sagging or flabby. Flank thin; tucked up or drawn in.

11. HAMS AND RUMP—Hams broad; full, long and wide. They should be as wide at point of the hip as at the swell of the ham. Buttocks large and full; should project beyond and come down upon and fill full between the hocks. The lower front part of the ham should be full, and stifle well covered with flesh, and a gradual rounding towards the hock. Rump should have a rounding slope from loin to root of tail; same width as back, and filling out full on each side of, and above the tail, ... 10

Objections—Ham narrow; short; thin; not projecting beyond and coming down to hock; cut up too high in crotch or twist; lacking in fullness at top or bottom; lacking in width from stifle straight back; lower fore part thin and flat; straight from root of tail to hock; buttocks light, thin or flabby. Rump flat, narrow and peaked at root of tail; too steep.

12. LEGS AND FEET—Legs medium length; straight; set well apart and squarely under body; tapering; well muscled and wide above knee and hock; below hock and knee round and tapering, capable of sustaining weight of animal in full flesh without breaking down; bone firm and of fine texture; pasterns short and nearly upright. Feet firm; short; tough and free from defects, ... 10

    Objections—Legs long; slim; coarse; crooked; muscles small above hock and knee; bone large, coarse; as large at foot as above knee; pasterns long, slim, crooked or weak; the hocks turned in or out of straight line; legs too close together; hoofs long, slim and weak; toes spreading or crooked, or unable to bear up weight of animal without breaking down.

13. TAIL—Well set on; small, smooth, tapering, and carried in a curl, ... 1

    Objections—Coarse; long; crooked, or hanging straight down like a rope.

14. COAT—Fine; straight; smooth; laying close to and covering the body well; not clipped; evenly distributed over body, ... 2

    Objections—Bristles; hair coarse; harsh; thin; wavy or curly; swirls; standing up; ends of hair split and brown; not evenly distributed over all of the body except belly. Clipped coats should be cut 1.5 points.

15. COLOR—Black, with white in face or on lower jaw; white on feet and tip of tail, and a few small, clear white spots on body not objectionable, ... 2

    Objections—Solid black, more than one fourth white; sandy hairs or spots; a grizzled or speckled appearance.

16. SIZE—Large for age and condition; boars two years old and over, if in good flesh, should not weigh less than 500 pounds. Sows same age and condition, not less than 450 pounds. Boars eighteen months old, in good condition, not less than 400 pounds; sows, 350 pounds. Boars twelve months, not less than 300 pounds; sows, 300 pounds.

Boar and sows, six months, not less than 150 pounds. Other ages in proportion, ... 5

Objections—Overgrown; coarse; gangling, or hard to fatten at any age.

17. ACTION and STYLE—Action vigorous; easy; quick and graceful. Style, attractive; high carriage; and in males, testicles should be of same size, carriage, readily seen, and yet not too large, ... 4

Objections—Slow; dull; clumsy; awkward; difficulty in getting up when down; low carriage; wabbling walk. In males, testicles not easily seen, not of same size or carriage, too large or only one showing.

18. CONDITION—Healthy; skin clear of scurf, scales, or sores; soft and mellow to the touch; flesh fine, evenly laid on and free from lumps or wrinkles. Hair soft and lying close to body; good feeding qualities, ... 4

Objections—Unhealthy; skin scaly, wrinkly, scabby or harsh; flabbiness or lumpy flesh; too much fat for breeding. Hair harsh, dry and standing up from body; poor feeders; deafness, partial or total.

19. DISPOSITION—Quiet and gentle and easily handled, . . 3

Objections—Cross, restless, vicious or wild.

PERFECTION, - - . - 100

SMALL YORKSHIRE SWINE.

Standard of Excellence for Small Yorkshire Swine, as adopted by the Small Yorkshire Club, G. W. Harris, Secretary, 3410 Third Avenue, New York; also by the American Yorkshire Club, W. F. Wilcox, Secretary, 148 Highland Avenue, Minneapolis, Minn. [Slightly changed in arrangement for this publication.]

SCALE OF POINTS FOR SMALL YORKSHIRE SWINE.

POINTS.	COUNTS
A. Head,	15
B. Trunk,	30
C. Hams,	25
D. Shoulders,	10
E. Legs,	5
F. Skin,	5
G. Hair,	5
H. General Appearance,	5
PERFECTION,	100

DETAILED DESCRIPTION.

POINTS.　　　　　　　　　　　　　　　　　　　　　COUNTS.

HEAD—Fifteen Points.

1. SMALLER THE BETTER, ... 2
2. NOSE—Shorter the better, ... 5
3. DISH—Greater the better, ... 3
4. WIDTH BETWEEN EARS—Greater the better, . . . 3
5. EARS SMALL, THIN, ERECT—More so the better, and may be pricked forward, but not lopped, ... 2

TRUNK—Thirty Points.

6. TOP LINE—Straighter the better, from shoulder to tail. . 5
7. BELLY LINE—The more level the better, . . . 5
8. GIRTH IN EXCESS OF LENGTH—More the better, if not more than ten per cent., ... 5
9. DEPTH—Greater the better, ... 5
10. WIDTH—Greater and evener the better, from shoulder to ham, ... 5
11. LOIN—Broader the better, ... 3
12. FLANK—Deeper and fuller the better, . . . 2

HAMS—Twenty-five Points.

13. Length—Longer the better, ... 10
14. Breadth—Broader the better, ... 10
15. Thickness—Greater the better, ... 5

SHOULDERS—Ten Points.

16. Length—Longer the better, ... 5
17. Breadth—Broader the better, ... 5
18. Thickness— 0

LEGS—Five Points.

19. Shorter the better, ... 3
20. Straighter the better, ... 2
21. Skin—Smooth, flexible, fine—more so the better. Must not be too thin nor ridgy and coarse, nor show discolored spots from old sores; not pale and ashy, but healthy in color and free from eruption, ... 5
22. Hair—Evener, finer, and thicker the better, ... 5
23. General Appearance—Symmetry and evidence of vigorous health, ... 5

Perfection, - - - - 100

DISCOUNTS AND DISQUALIFICATIONS.

DISCOUNTS.

1. Pedigree—Lack of registration or eligibility to be registered disqualifies, ... 100 points.
2. Sterility—Inability to produce offspring disqualifies, ... 100 "
3. Deformity—Any structural deformity or lack disqualifies, ... 100 "
4. Disease—Any evidence of, or tendency to disease, disqualifies, ... 100 "
   " Scars of sores, discolored spots, eruptions, excema; etc., ... 5 to 25 "
5. Colored Hair—Disqualifies, ... 100 "
6. Colored Spots—Dark spots in skin, . . . 5 to 25 "
7. Size—Inordinate size, with coarseness of bone or form, ... 10 to 50 "
8. " Diminutive size, ... 5 to 25 "
9. Disposition—Savage or fierce nature, . . 5 to 10 "

SUFFOLK SWINE.

Standard of Excellence for Suffolk Swine, as adopted by the American Suffolk Association, W. F. Watson, Secretary, Winchester, Indiana.

SCALE OF POINTS FOR SUFFOLK SWINE.

POINTS.	COUNTS.
1. Color—White,	2
2. Head—Small, broad and face dished,	3
3. Ears—Fine, erect, slightly drooping with age.	2
4. Jowl—Full and neat.	1
5. Neck—Short, full and slightly arched,	3
6. Shoulders—Broad and deep,	7
7. Girth around heart,	6
8. Back—Straight, broad and level,	12
9. Sides—Deep and full,	6
10. Ribs—Well sprung,	7
11. Loin—Broad and strong,	12
12. Flank—Well let down,	2
13. Ham—Broad, full and deep,	12
14. Tail—Medium, fine and curled,	2
15. Legs—Fine, straight and tapering,	3
16. Feet—Small,	3
17. Hair—Fine and silky, free from bristles,	3
18. Action—Easy and graceful,	4
19. Symmetry—Adaptation of the several parts to each other,	10
PERFECTION,	100

TAMWORTH SWINE.

Quite recently many specimens of the Tamworth breed of swine have been imported from England into the United States and Canada. At the time this book goes to press the Tamworth breeders have failed to organize an association of their own, although an attempt was made in Massachusetts. The following description from the Breeders' Gazette, October 18th, 1893, gives an excellent idea of the breed, especially as they appeared at the World's Fair, at Chicago:

"If the caricaturist were asked to describe this big, sandy-haired breed of swine as most of them appeared at the Columbian he would probably depict them as all snouts and slab-sides. And the caricature would be so near the truth as almost to miss being a caricature. As a matter of fact the Tamworths shown from Canada were the sensation of the swine show. They were *sui generis* and so suggestive in their conformation of the thoroughbred "hazel-splitter" as to be the butt of ridicule throughout the showing. It is said that the bacon curers of Canada are strongly urging these swine upon the Dominion pig-breeders. We can readily believe it, for the bacon-curer cares nothing for the hams or shoulders, and the Tamworths have little of either. He wants sides, and the Tams are literally "long" on sides, and deep also. They are tremendously high and deep-sided, remarkably light in hams and shoulders, and prodigiously prodigal of snout and ears. Lean side meat they grow in great quantity, but if it approaches in quality the finer-grained breeds our ideas of form as related to quality of flesh need revision. That such long-nosed, slab-sided swine can be easy feeders is a proposition which no amount of argument could force upon the grower of pigs for the American markets. Mr. Thomas Bennett, Rossville, Ill., was showing an entirely different type of Tamworths. Ten years ago he personally made an importation of this breed, selecting a shorter-legged, wider-backed, more compact type, and he has been improving them ever since, having imported another boar for this purpose about two years ago. The type he now shows is that most in favor in American feed-lots, and as the judge, Mr. F. D. Coburn, was looking for the kind which more nearly meets modern ideas in pork-making, Mr. Bennett's pigs came in for chief recognition. The story of the showing is a brief one. Exhibitors were Thomas Bennett, Rossville, Ill.; James Calvert, Thedford, Ont., and John Bell, Amber, Ont."

VICTORIA SWINE.

Detailed description of Victoria Swine, as adopted by the Victoria Swine Breeders' Association, George F. Davis, Secretary, Dyer, Indiana, at their annual meeting, November, 1888, as an aid to Judges at Fairs, in place of the score card, and to assist breeders to establish uniformity.

DISQUALIFICATIONS.

COLOR: Other than white or creamy white, with occasional dark spots in skin.

FORM: Crooked jaws or deformed face; crooked or deformed legs; large, coarse, drooping ears.

CONDITION: Excessive fatness; barrenness; deformity in any part of the body.

PEDIGREE: Not eligible for record.

POINTS. **DETAILED DESCRIPTION.**

1. COLOR—White, with occasional dark spots in the skin.
2. HEAD AND FACE—Head rather small and neat. Face medium dished and smooth; wide between eyes; tapering from eyes to nose.
3. EYES—Medium size, prominent, bright; clear and lively in young and quiet expression in aged animals.
4. EARS—Small, thin, fine, silky; upright in young pigs, pointing forward and slightly outward in aged animals.
5. NECK—Medium wide, deep, short, well arched and full at top.
6. JOWL—Medium full, nicely rounded, neat and free from loose, flabby fat.
7. SHOULDERS—Broad, deep and full, not higher than line of back, and as wide as top of back.
8. CHEST—Large, wide, deep and roomy, with large girth back of shoulders.
9. BACK AND LOIN—Broad, straight or slightly arched, carrying same width from shoulder to ham; level and full at loin; sometimes slightly higher at hip than at shoulders.
10. RIBS AND SIDES—Ribs well sprung at top; strong and firm; sides deep, full, smooth and firm; free from creases.
11. BELLY AND FLANK—Wide; straight and full; as low or slightly

lower at flank than at chest. Flank full and nearly even with sides.

12. HAMS AND RUMP—Hams long; full and wide; nicely rounded; trim and free from loose fat. Buttocks large and full; reaching well down to hocks. Rump slightly sloped from end of loin to root of tail.

13. LEGS AND FEET—Legs short; set well apart and firm; wide above knee and hock tapering below. Feet firm and standing well up on toes.

14. TAIL—Small; fine and tapering; nicely curled.

15. COAT—Fine and silky; evenly covering the body.

16. SIZE—Boars two years old and over when in good condition should weigh not less than 500 pounds; sow same age and condition, 450 pounds. Boars twelve months old not less than 300 pounds; sows in good flesh 300 pounds. Pigs five to six months old 140 to 160 pounds.

17. ACTION—Easy and graceful but quiet.

18. CONDITION—Healthy; skin clean, and white or pink in color; free from scurf; flesh firm and evenly laid on.

19. DISPOSITION—Quiet and gentle.

APPENDIX.

SHETLAND PONIES.

Standard of Excellence for Shetland Ponies, as adopted by the American Shetland Pony Club, Mortimer Levering, Secretary, Lafayette, Indiana.

SCALE OF POINTS FOR SHETLAND PONIES.

POINTS.	COUNTS
1. Constitution,	10
2. Size,	25
3. Head,	10
4. Body,	10
5. Legs,	25
6. Mane and Tail,	10
7. Feet,	10
PERFECTION,	100

DETAILED DESCRIPTION.

POINTS.   COUNTS.

1. CONSTITUTION—Constitution indicated by general healthy appearance, perfect respiration, brightness of eyes, . . 10
2. SIZE—Ponies over four years old, 42 inches and under in height; two points to be deducted for every inch over 42 inches up to 46 inches, fractional portions to count as full inches. Ponies over 46 inches in height ineligible to registry, ... 25
3. HEAD—Head symmetrical, size proportionate to body, wide between the eyes, ears short and erect, jaw full and deep, 10
4. BODY—Barrel well rounded, back short and level, deep chested, good breast, compact, "pony build" . . 10
5. LEGS—Legs muscular, flat-boned, hind legs not cow-hocked or too crooked, ... 25
6. MANE AND TAIL—Foretop, mane and tail heavy, . . 10
7. FEET—Good, ... 10

                           PERFECTION, - - - 100

MEASUREMENTS REQUIRED.

Height at withers in line with foreleg, ... Inches
Measurement of the girth around heart, ... Inches
Weight, ... Pounds

N. B.—The following letter will be found to contain several valuable suggestions in regard to judging Shetland ponies:

AMERICAN SHETLAND PONY CLUB.

Secretary's Office,
La Fayette, Ind., April 18, 1893.

Frank A. Lovelock, Esq.:

Dear Sir:—In answer to your favor of 10th instant will say, the maximum height of ponies, as you will see by the rules, is 46 inches. The smaller the pony, if he is blocky and well formed, the higher he will be considered in class. Some of the finest Shetlands in the country, and the highest priced are from 28 to 31 inches high, and weigh about 200 pounds. The type of Shetland is as far from the thoroughbred horse type as possible to get. The pony must be square-built, strong, large limbs, small head and ears. All ponies on the thoroughbred type are undesirable. It is said the pure-bred Shetland cannot kick over eight inches from the ground. They are intended for children's pets, and we do not want to type a pony that can stand and kick a man's hat off, as can some of the Welch Exmoor ponies. Ponies of 44 to 45 inches high should weigh from 350 to 400 pounds.

Yours respectfully,
Mortimer Levering,
Secretary.

SHORT-HORN CATTLE.

A few days before this book went into the hands of the binder, the editor found on the agricultural page of the Philadelphia Public Ledger, the following standard of excellence for Short-Horn Cattle, as established by the Massachusetts State Board of Agriculture, for use at the Fairs of that State. Without wishing to criticise, the editor would say that he prefers for his own use the standard of excellence for Short-Horn Cattle as found on pages 44, 45 and 46 of this work.

SCALE OF POINTS FOR SHORT-HORN COW.

POINTS.	COUNTS.
1. Head,	3
2. Face,	2
3. Eye,	2
4. Horns,	1
5. Neck,	2
6. Chest,	14
7. Brisket,	5
8. Shoulder,	4
9. Crops,	8
10. Back, Loin and Hips,	8
11. Rump,	5
12. Pelvis,	2
13. Twist,	3
14. Quarters,	5
15. Carcass,	4
16. Flanks,	3
17. Legs,	2
18. Plates of the Belly,	3
19. Tail.	2
20. Carriage,	2
21. Quality,	15
22. Coat,	2
23. Udder,	3
PERFECTION,	100

DETAILED DESCRIPTION.

POINTS. COUNTS.

1. HEAD—Small, lean and bony, tapering to the muzzle, . . 3
2. FACE—Somewhat long, the fleshy portion of the nose of a light, delicate color, ... 2
3. EYES—Of great significance, and should be prominent,

bright and clear—"prominent" from an accumulation of "adeps" in the back part of its socket, which indicates a tendency to lay on fat, "bright" as an evidence of a good disposition, "clear" as a guaranty of the animal's health; whereas a dull, sluggish eye belongs to a slow feeder, and a wild, restless eye betrays an unquiet, fitful temper, . . 2

4. HORNS—Light in substance and waxy in color, and symmetrically set on the head; the ear large, thin, and with considerable action, ... 1

5. NECK—Rather short than long, tapering to the head, clean in the throat, and full at its base, thus covering and filling out the points of the shoulders, ... 2

6. CHEST—Broad from point to point of the shoulders, deep from the anterior dorsal vertebra to the floor of the sternum, and both round and full just back of the elbows, sometimes designated by the phrase "thick through the heart." These are unquestionably the most important points in every animal, as constitution must depend on their perfect development, and the ample room thus afforded for the free action of the heart and lungs, . . 14

7. BRISKET—However deep or projecting, must not be confounded with capacity of chest, for though a very attractive and selling point, it, in reality, adds nothing to the space within, however it may increase the girth without. It is, in fact, nothing more nor less than a muscular adipose substance, attached to the anterior portion of the sternum, or breast-bone, and thence extending itself back. This form, however, of the brisket indicates a disposition to lay on fat generally throughout the frame, and in this point of view is valuable, ... 5

8. SHOULDER—Where weight, as in the Shorthorn, is the object, should be somewhat upright and of good width at the points, with the blade-bone just sufficiently curved to blend its upper portion smoothly with the crops, . . 4

9. CROPS—Must be full and level with the shoulders and back, and is, perhaps, one of the most difficult points to breed right in a Shorthorn, ... 8

10. BACK, LOIN AND HIPS—Should be broad and wide, forming a straight and even line from the neck to the setting on of the tail, the hips or hooks round and well covered, . . 8

11. RUMPS—Laid up high, with plenty of flesh on their extremities, ... 5
12. PELVIS—Should be large, indicated by the width of the hips (as already mentioned) and the breadth of the twist, . . 2
13. TWIST—Should be so well filled out in its "seam" as to form nearly an even and wide plain between the thighs, . . 3
14. QUARTERS—Long, straight and well developed downwards, 5
15. CARCASS—Round, the ribs nearly circular and extending well back, ... 4
16. FLANKS—Deep, wide and full in proportion to condition, . 3
17. LEGS—Short, straight and standing square with the body, 2
18. PLATES—Of the belly strong, and thus preserving nearly a straight underline, ... 3
19. TAIL—Flat and broad at its root, but fine in its cord, and placed high up and on a level with the rumps, . . . 2
20. CARRIAGE—Of an animal gives style and beauty; the walk should be square and the step quick, the head up, . . 2
21. QUALITY—On this the thriftness, the feeding properties and the value of the animal depend; and upon the touch of this quality rests, in a good measure, the grazier's and the butcher's judgment. If the "touch" be good, some deficiency of form may be excused; but if it be hard and stiff, nothing can compensate for so unpromising a feature. In raising the skin from the body, between the thumb and finger, it should have a soft, flexible and substantial feel, and when beneath the outspread hand it should move easily with it, and under it, as though resting on a soft, elastic, cellular substance; which, however, becomes firmer as the animal ripens. A thin papery skin is objectionable, more especially in a cold climate, ... 15
22. COAT—Should be thick, short and mossy, with longer hair in winter, fine, soft and glossy in summer, ... 2
23. UDDER—Pliable and thin in its texture, reaching well forward, roomy behind, and the teats standing wide apart, and of convenient size, ... 3

PERFECTION, . . . 100

THE BULL.

The points desirable in the females are generally so in the male, but must, of course, be attended by that masculine character which is inseparable from a strong, vigorous constitution. Even a certain degree of coarseness is admissible, but then it must be so exclusively of a masculine description as never to be discovered in the females of his get.

In contradistinction to the cow, the head of the bull may be shorter, the frontal bone broader, and the occipital flat and stronger, that it may receive and sustain the horn, and this latter may be excused if a little heavy at the base, so its upward form, its quality and color be right. Neither is the looseness of the skin attached to and depending from the under jaw to be deemed other than a feature of the sex, provided it is not extended beyond the bone, but leaves the gullet and throat clean and free from dewlap.

The upper portion of the neck should be full and muscular, for it is an indication of strength, power and constitution. The spine should be strong, the bones of the loin long and broad, and the whole muscular system wide and thoroughly developed over the entire frame.

TUNIS OR BROAD-TAILED SHEEP.

Standard of Excellence for Tunis or Broad-Tailed Sheep, as adopted by the American Tunis Sheep Breeders' Association. G. A. Guilliams, President, and M A. Bridges, Secretary, Fincastle, Indiana. [This standard was sent to the editor just as this book was ready for the binder, hence could not appear in the sheep department.]

SCALE OF POINTS FOR TUNIS OR BROAD-TAILED SHEEP.

POINTS.	COUNTS.
1. Blood,	20
2. Constitution,	15
3. Fleece,	10
4. Covering,	10
5. Form and Tail,	12
6. Head and Ears,	10
7. Neck,	5
8. Legs,	6
9. Size,	6
10. General Appearance,	6
PERFECTION,	100

DETAILED DESCRIPTION.

POINTS. COUNTS

1. BLOOD—Imported from Tunis, or having a perfect line of ancestors extending back to the flock owned and bred by Judge Richard Peters, of Belmont, near Philadelphia, who received his first pair from Tunis in 1779, and bred them pure for more than 20 years, ... 20

2. CONSTITUTION—Healthful countenance, lively look, head erect, deep chest, ribs well arched, round body with good length. Strong, straight back; muscles fine and firm, . 15

3. FLEECE—Medium length, medium quality, medium quantity, color tinctured with gray, never pure white. Evenness throughout, ... 10

4. COVERING—Body and neck well covered with wool. Legs bare or slightly covered; face free from wool and covered with fine hair, ... 10

5. FORM AND TAIL—Body straight, broad and well propor-

tioned. Small bone; breast wide and prominent in front. Tail—the little end should be docked, leaving the fleshy part fan shaped, or tapering; five to ten inches broad, six or eight inches long and well covered with wool, . . 12

6. HEAD AND EARS—Head small and hornless, tapering to end of nose; face and nose clean, in color brown and white. Ears broad, thin, pendulous, covered with fine hair; in color brown to light fawn, ... 10

7. NECK—Medium in length, well placed on shoulders; small and tapering, ... 5

8. LEGS—Short. In color, brown and white; slightly wooled not objectionable, ... 6

9. SIZE—In fair condition, when fully matured, rams should weigh 150 pounds and upwards; ewes, 120 pounds and upwards, ... 6

10. GENERAL APPEARANCE—Good carriage; head well up; quick elastic movements, showing symmetry of form and uniformity of character throughout, ... 6

PERFECTION, - - - 100

CONTENTS.

N.

Nomenclature for Bull,	8
Nomenclature for Cow,	9
Nomenclature for Hog,	98

O.

Oxford Down Sheep,	88, 89

P.

Poland China Swine,	118 125
Polled Durham Cattle,	46
Preface,	3, 4

R.

Red Polled Cattle,	42, 43
Reform in Judging at Fairs,	5 7

S.

Shetland Ponies,	133, 134
Short-horn Cattle,	44 46; 135 138
Shropshire-down Sheep,	90, 91
Small-Yorkshire Swine,	126, 127
Southdown Sheep,	92, 93
Suffolk Sheep,	94, 95
Suffolk Swine,	128
Sussex Cattle,	47, 48
Swiss (Brown) Cattle,	49 52

T.

Tamworth Swine,	129
Tunis or Broad Tailed Sheep,	139, 140

V.

Victoria Swine,	130, 131

W.

West Highland Cattle,	53, 54

www.ingramcontent.com/pod-product-compliance
Lightning Source LLC
Chambersburg PA
CBHW062216220526
45471CB00009B/3221